Social Statistics

Third Edition

William Fox
Skidmore College

MicroCase Corporation
Bellevue, Washington

Editor	David Smetters
Editorial Assistant	Julie Aguilar
Production Manager	Jodi Gleason
Cover/Interior Design	Michael Brugman Design
Copy Editor	Margaret Moore

Windows 95 is a registered trademark of Microsoft Corporation.

MicroCase and **ExplorIt** are registered trademarks of MicroCase Corporation.

Printed in the United States of America

1 2 3 4 5 6 7 8 9 10–01 00 99 98

To Collette and John and Jennifer,
for all that is most important.
Family matters.

Contents

Part I: Introduction and Univariate Analyses

Part III: Multivariate Analyses

Appendix: Statistical Tables

About the Author

William Fox has a B.A. in Political Science from the University of Michigan and a Ph.D. in Sociology from Indiana University. He has taught at the University of Iowa, Florida Atlantic University, and Richmond College in London, and is now Professor of Sociology at Skidmore College in upstate New York. His current research interests are in urban folklore and social history. Bill lives with his wife, Collette, in an empty Victorian nest in Saratoga Springs, New York, where he enjoys biographies, Mozart, and country music.

Preface

This text uses a student version of MicroCase, a straightforward, no nonsense piece of statistics software. MicroCase is written to do data analysis and get the job done efficiently, competently, and intelligently. Students learn statistics—and substantive social science too, for that matter—with minimal distraction from the computer itself. No inundation with esoteric options. Just learning statistics (or whatever). Just doing data analysis. I have written this text in the same pragmatic spirit as MicroCase.

This third edition is prompted by the generous reception given previous editions and is guided by their users' comments and suggestions. Students and faculty alike have forwarded their ideas for conveying statistical ideas more effectively and efficiently. I have tried my best to incorporate the sense and substance of their suggestions in this edition. Thus I have added a chapter on the t test of the difference between means and expanded discussions of confidence intervals and beta coefficients. This edition uses updated and expanded data sets and incorporates entirely new versions of the MicroCase software, including a version for Windows 95 users. I have integrated data management techniques into statistical analysis and offered examples of statistical write-ups. Students welcomed the supportive, interactive style of earlier editions, and I have continued the same in this edition.

What You Need to Bring to This Text

You are probably interested in what you need to know *before* you begin to learn statistics. I assume you know basic arithmetic—adding, subtracting, multiplying, and dividing. You should also be reasonably comfortable with fractions and decimals, squares and square roots, and simple equations.

Whatever else you need to know, you will learn along the way. All the better if you have had a course in research methods, but that is not essential. Likewise, I hope you have had several substantive courses in the social sciences, but those too are not really necessary. Frankly, I assume more motivation and good will on your part than I do prior knowledge.

I trust that you have a good statistics teacher. I think good textbooks matter (that's why I have worked hard on this one), but reading even the best text is less important than learning from a knowledgeable and enthusiastic teacher.

Oh yes, you also need a calculator. An inexpensive one is fine as long as it finds square roots. Some graph paper will also be useful.

Acknowledgments

This third edition was written while on sabbatical leave from Skidmore College. I am fortunate indeed to teach at a college so committed to the liberal arts, high-quality teaching, and support of its faculty . . . and so cognizant that these three commitments are one.

This text began with my former colleague Richard Rosenfeld, now at the University of Missouri at St. Louis. Rick extolled the virtues of MicroCase, gently nagged me to try it, suggested the need for an accompanying stat text, and then actively supported my project at critical points. Without Rick I would not have published this book.

Many users of previous editions offered suggestions that were most helpful for this revision. Especially valuable have been comments and ideas from these colleagues across the country:

Gretchen Arnold, University of Missouri - St. Louis
Scott Beck, East Tennessee State University
James C. Cramer, University of California - Davis
Thom Curtis, University of Hawaii - West Hawaii
Robert Evans, Rockford College
Jody L. Fitzpatrick, University of Colorado
George Guay, Salem State College
Farrukh B. Hakeem, John Jay College of Criminal Justice
Craig C. Hagensick, University of Wisconsin - Milwaukee
Frances Hoffmann, University of Missouri - St. Louis
Rosemary Hopcroft, University of North Carolina - Charlotte
Rollie Jacobs, U.S. Military Academy
Michael Jerin, St. Ambrose University
Jocelyn Johnston, University of Kansas
Michael C. Kearl, Trinity University
Scott Kuehn, Clarion University of Pennsylvania
Gary Lilly, Westminster College

Mary Maloney, Wayland Baptist University
William F. McDonald, Georgetown University
Miller McPherson, University of Arizona
Kenneth Mietus, Western Illinois University
Carol Miller, Humboldt State University
David A. Nordlie, Bemidji State University
Jeff Pounders, Ouachita Baptist University
John W. Prehn, Gustavus Adolphus College
Roger Reed, Dakota State University
Steve Roberds, University of North Alabama
Beth Rushing, University of Missouri at St. Louis
Terry Russell, Frostburg State University
Richard Shingles, Virginia Polytechnic Institute & State University
Kurt Schock, Rutgers University - Newark
Christopher B. Smith, Mount Saint Mary's College
Kurt Thurmaier, University of Kansas
Rollo Tinkler, Abilene Christian University
Henry Vandenburgh, Prairie View A&M University
Mark Wattier, Murray State University
Raymond Wheeler, University of South Florida
James Wiest, Hastings College
Sheldon Zhang, California State University - San Marcos

Special thanks to Steve Bird and Eric Metchik for so carefully reviewing later drafts of this text and the accompanying workbook. Any errors that remain are mine alone.

The MicroCase folks—Dave Smetters, David Simmons, Julie Aguilar, and Jodi Gleason—eagerly welcomed this book and did everything they could to aid its progress. They developed Student MicroCase, offered advice, handled critiques of drafts, and provided MicroCase displays. Most important of all, however, they offered enthusiasm. I owe special thanks to Dave Smetters for seeing this project through with an optimal blend of professionalism and empathy.

Thanks to my students over the years, especially at Skidmore College. True, they couldn't stop me from trying different teaching strategies on them. But they need not have been so good-natured nor have taught me so much, and for that I am grateful. I received all sorts of assistance from Skidmore College's Computer specialists, especially Leo Geoffrion, John Danison, Lisa Schermerhorn, and Bill Duffy. These are the user-friendliest bunch of alpha geeks anywhere. Susan Danielson, my assistant, worked on this textbook project with extraordinary diligence and moxie.

I want to thank Elton Jackson, Professor of Sociology at Indiana University, for teaching me statistics. On the first day of class, Elton promised a new way of seeing the world. He delivered on that promise. I learned statistics and much more from Elton Jackson.

My appreciation goes to several individuals and institutions who provided the data files upon which most of the workbook exercises are based. First, I obtained all my data sets from MicroCase's Data Archive, which saved me weeks of work. Special thanks goes to Tom W. Smith at the National Opinion Research Center for his continued direction and administration of the General Social Survey. Thanks also goes to the Inter-University Consortium for Political and Social Research (ICPSR) and to The Roper Center for their many years of distribution of the General Social Survey. The data archives maintained by ICPSR, The Roper Center, and MicroCase Corporation are an invaluable service to social science researchers.

I am grateful to the Literary Executor of the late Sir Ronald A. Fisher, F.R.S., Dr. Frank Yates, F.R.S., and the Longman Group Ltd, London, for permission to reprint parts of Tables III and IV from their book *Statistical Tables for Biological, Agricultural and Medical Research*.

But most of all, I want to thank Nanci Griffith and Collette Fox. Nanci and all her acoustic pals make great music that goes so well with statistics and just about everything else in life. And Collette makes that life worthwhile.

Any Comments or Suggestions?

I want very much to hear from you if you have any helpful feedback concerning this text. Let me know what you like and don't like about it, what works well and what can be improved, what might be left out and what needs to be added. If you have Internet access, write me at statprof@skidmore.edu. Or you can write me by snail mail at the following address:

William Fox
Department of Sociology, Anthropology,
    and Social Work
Skidmore College
Saratoga Springs, NY 12866

One final word before we start learning statistics: I had a great time writing this book. I hope you enjoy it just as much.

Bill Fox
Skidmore College

PART

I

Introduction and Univariate Analyses

CHAPTER 1
Statistics and Variables

In this introductory chapter I will describe what statistics is, what variables are, and what data looks like. I will then offer an overview of how we will go about learning statistics using a friendly computer program called Student MicroCase to analyze interesting sets of data. The *Doing Statistics Using MicroCase* workbook will then get you started using MicroCase and offer exercises that will help you learn statistics.

After this chapter you will be able to:

1. Explain the difference between descriptive and inferential statistics.

2. Explain the difference between a sample and a population.

3. Distinguish between a statistic and a parameter.

4. Explain what a data set and a data file are.

5. Recognize units of analysis and cases.

6. Recognize a variable and offer examples of variables.

7. Explain what a dichotomous variable is and offer examples of dichotomous variables.

8. Distinguish between continuous and discrete variables and offer examples of each.

9. Distinguish among nominal, ordinal, and interval/ratio levels of measurement, offer examples of each, and identify levels of measurement of variables.

10. Explain what aggregate and ecological data are and offer examples of each.

11. Appreciate *thinking* as the most important process in statistics.

1.1   Statistics and Data

Sometimes we use the word *statistics* to refer to numbers that summarize information quantitatively. We say that baseball player Joe DiMaggio had "good stats" with his lifetime batting average of .325, or that statistics show that violent crime decreased 3.6 percent last year, or that the average daily attendance at Yankee Stadium was 32,505 per game last summer, or that 54 percent of Americans surveyed think the President is doing a good job. These sorts of numerical summaries are indeed statistics and we encounter them often. Modern societies float in a sea of statistics. Although we often complain about statistical information ("They're trying to reduce everything to a number!"), we know we could not function very well without statistics. We need quantitative summaries—statistics—to understand the world clearly and to make decisions intelligently.

But *statistics* also means the *methods* used to calculate these summary numbers and to generalize from them. In this sense, statistics refers to ways we make information more manageable—ways to calculate batting averages or describe crime trends or determine average attendance or generalize from a sample of Americans to the entire electorate. In this text I will use the word *statistics* mainly in this second sense to mean methods for quantitatively summarizing and generalizing information.[1]

The unsummarized, "raw" information that statistics (in this second sense) makes more manageable is called *data*.[2] Data are records of observations. Most data that social scientists work with come from experiments or surveys, but data can come from systematic observations of any kind. For example, official baseball scorers recorded what happened each time Joe DiMaggio batted. The record of all DiMaggio's 6821 times at bat is data that statistics then summarizes as a batting average.

To make information easier to use, raw data from observations are organized in some systematic way—for example, as a carefully compiled, uniform listing of Joe DiMaggio's batting records. We call data in this organized form a *data set*. Data sets stored so that they can be read by a computer are called *data files*. Sometimes data files also include descriptions of the data along with the raw data. Several data files are stored on the computer diskette accompanying this text's workbook. In this text we will draw on information from three data sets—one from a large survey of adult Americans (the General Social

---

[1] A little later in this chapter we will find that *statistics* is also used in a narrow technical sense to mean characteristics of samples as opposed to entire populations.

[2] It is OK to use "data" as either a singular or plural noun as long as you are consistent. It is also OK to say either dā'ta or dät'a—like tomāto or tomäto.

Survey), another for the 50 American states, and a third for the world's 50 most populous nations.

A *unit of analysis* is the person, object, or event that a researcher is studying. Here are some of the many units of analysis that social scientists use: individual persons, groups, families, colleges, countries, cultures, newspaper editorials, presidential elections, and wars. Although individual people and groups of people are surely the most common unit of analysis for social scientists, note that units of analysis can be things such as editorials and events such as elections. A *case* is the specific unit from which data are collected. So, for example, individuals surveyed by the Gallup Poll are cases. So too are countries included in a data set for the 50 largest nations, or states in a data set for the American states, or editorials included in a study of newspapers, or presidential elections in a political data set.

1.2 Overview of Statistics

There are two major branches of statistics—descriptive and inferential—that correspond respectively to summarizing and generalizing information. *Descriptive statistics* refers to methods for summarizing information so that the information is more intelligible or more useful or can be communicated more effectively. Examples of descriptive statistics are methods for calculating averages and graphing techniques for displaying information visually. Descriptive statistics can also describe relationships among variables, as, for example, many percentage tables do.

Descriptive statistics always involves some loss of detail but if applied sensibly can make information far more useful. Although Joe DiMaggio's .325 lifetime batting average does not describe what Joltin' Joe did each of his 6821 times at bat, it does give an overall summary of how good a batter he was. (He was very good indeed.) The ups and downs of monthly crime statistics over many years may be confusing, but a summary describing general trends is extremely helpful in formulating public policy. Attendance at Yankee Stadium varies so widely from game to game that we cannot make much sense of detailed daily information, but average daily attendance provides an intelligible summary of the Yankees' popularity last summer. And it is difficult to make much sense of 1500 individual opinions about the president without summarizing them into a more general index such as the percentage of people polled who think the president is doing a good job.

If we were omniscient beings with immense minds (like *Jeopardy!* contestants), maybe we could do without descriptive statistics. Maybe we would handle all the detail just fine. But given the limitations of our human brains, we need some way to condense information into more manageable form. Descriptive statistics (devised by our limited but rather clever human brains) helps us reduce information to a form we can comprehend and use effectively.

Inferential statistics, the other major branch of statistics, refers to procedures used to generalize from a sample to the larger population and to assess the confidence we have in such generalizing. When a Gallup Poll finds that 54 percent of people interviewed think the president is doing a good job, the Gallup pollsters are not really interested in the particular 1500 people whom they happened to sample. After all, a different sample of 1500 persons would include different people with different opinions and thus might yield a different percentage—maybe 56 percent. And some other sample might find, say, 53 percent. And so on, with different samples yielding different percentages. No, the pollsters' interest is not in any particular sample. Their interest is in the larger population from which the sample was drawn—the entire American electorate. What can we say about the political opinions of *all* adult Americans if 54 percent of a sample think the President is doing a good job? And how much confidence can we have in our conclusions about the population? Pollsters use inferential statistics to answer these sorts of questions.

Just as Molière found himself speaking prose without knowing it, you have run into inferential statistics many times without labeling it as such. We rely on sample data so often that inferential statistics is common. Inferential statistics is used, for example, to find the margins of error often reported with poll results (e.g., "54 percent with a margin of error of ±3 percent"). Or maybe you have seen chi-square tests or expressions like "$p < .05$" in research you have read. These tests and expressions are inferential statistics used to generalize sample information to a population.

Descriptive statistics is more fundamental than inferential statistics in that we must summarize information about a sample before we can generalize the sample information to a larger population. But descriptive and inferential statistics complement one another, and both branches of statistics are essential for social scientific research. We will learn both descriptive and inferential statistics in this text.

1.3   Samples and Populations

Inferential statistics generalizes from a sample to a population, so we need to distinguish clearly between these two sources of data. Population data pertain to all or almost all cases to which a researcher wants to generalize. Data from censuses are population data. Despite its acknowledged incompleteness, the U.S. Census conducted every 10 years tries to include *everyone* in the United States. Election results, since they pertain to everyone who voted, also are population data. Research based on data for U.S. states or Canadian provinces uses population data, the population consisting of the 50 individual American states or the 12 Canadian provinces and territories. The world's 50 most populous nations are a population too. Whenever information is from all or nearly all cases that a researcher wants to describe, the data are population data.

Often, however, data from entire populations are too expensive, infeasible, or even impossible to collect. Imagine trying to conduct a *world* census. Or trying to analyze *all* telephone conversations. Or studying *every* pedestrian crossing *every* Chicago intersection. And if your doctor took a blood test using the population of *all* your blood cells, you would end up a bloodless corpse—an unpleasant prospect cutting short your research career. Furthermore, even if we can collect information from an entire population, often we can get more data and higher-quality data if we use a sample. Surveying, say, 1500 Americans allows us to ask far more questions, train interviewers much more carefully, and suffer far fewer data collection and processing errors than is the case with the U.S. Census.

That is why we so often rely on data from samples. We then generalize from the sample data to the population from which the sample was drawn. That, you'll recall, is what inferential statistics does. These generalizations can be made with much more confidence if we choose our samples very carefully using random sampling procedures. This is what the Gallup Poll and other good polling organizations do. They use very elaborate procedures for ensuring that the cases they select for their surveys are representative of the population to which they want to generalize. Random sampling is like picking cases out of a hat although quite sophisticated techniques are used instead of baseball caps or other headgear. Later in this chapter I will introduce you to the National Opinion Research Center (NORC) and its General Social Survey, one of the best social scientific data sets. NORC uses random sampling procedures in its General Social Survey.

The distinction between populations and samples is so important that statisticians use different words and symbols to refer to summaries of these two kinds of data. A *parameter* is a characteristic of a population; a *statistic* is a characteristic of a sample. So, for example, the average age, percent married, and average income of *all* Americans are parameters since they pertain to an entire population. The average age, percent married, and average income of a *sample* of Americans are statistics. In later chapters we will find that statisticians usually use Greek letters like σ to denote parameters and Roman letters like s to symbolize statistics.

1.4   Variables

Variables are so important in scientific research and everyday life that you already have a pretty good understanding of what a variable is. A *variable* is some characteristic or property that differs in value from one case to another. A characteristic is a variable if it has two or more values or attributes. Simply put: A variable is something that varies. The opposite of a variable is a constant—something that never varies, like pi or the speed of light or the half-life of plutonium. We don't have many constants in the social sciences. Mostly we have variables. Maybe that is why social science is so fascinating and complex.

An example of a variable is gender. Gender has two values: male and female. Another variable is social class, perhaps with the values lower, working, middle, and upper. Age is still another variable: a person can be 1 year old, 2 years old, 3 years . . . . Those are the values of the variable age. Other variables that we use often in social science are race, ethnicity, religion, education, income, alienation, authoritarianism, prejudice, political preference, number of children, region of the country . . . and on and on. These variables pertain to individuals. If we use countries as the unit of analysis, variables might include fertility rate, population density, percent literate, annual population growth . . . and, again, on and on. The list of variables is endless because we continue to create new variables to identify social patterns and to make sense of the world around us.

What makes a variable useful scientifically is measurement. *Measurement* is the process of finding the values of a variable for different cases. We can measure variables for any unit of analysis we study—individuals, families, groups, communities, entire societies, artifacts, and events. So, for example, we can measure an individual's education, a country's birth rate, a war's casualties.

We use the term *scale* to refer to a variable's set of values. Years of schooling is a scale for measuring a person's education. A country's birth rate can be measured as number of births per 1000 population. And casualties are measured with a scale of numbers of dead or wounded. The values of variables that we find when we measure something are called *scores*. Examples of scores on the three variables just mentioned are 16 years of school for Jodie Foster (B.A., Yale), 11 births per 1000 population for Germany, and 211,005 Americans dead or wounded in the Vietnam War.

Social scientists measure all sorts of things, including beliefs, opinions, and values. We even measure people's positions on moral or ethical issues. The large General Social Survey that I'll describe later in this chapter asked respondents, "In general, would you say that people should obey the law without exception or are there exceptional occasions on which people should follow their consciences even if it means breaking the law?" Response choices were obey the law and follow one's conscience. The issues raised by this survey question are profound: the relationship of the individual to the community; the limits of law; the morality of civil disobedience. Socrates wrestled with these issues, choosing to drink hemlock in obedience to the laws of Athens. Henry David Thoreau, refusing to pay his tax to protest his government's war with Mexico, and Martin Luther King, Jr., going to jail to protest the injustices of racial segregation, faced these issues. Obviously a survey question cannot explore the deep philosophical and political implications of obeying the law versus following one's conscience, but responses do describe variations in the public's general stance toward civil disobedience. Each survey respondent's answer to this question—obey the law or follow one's conscience—reflects his/her position with regard to civil disobedience.

1.5   Levels of Measurement

It is helpful to classify variables into four types based on how they are measured. The types are nominal, ordinal, interval, and ratio variables. Each of these four categories, called a *level of measurement*, describes how much information is conveyed by differences between values of the variable. Levels of measurement are important for deciding how to analyze data statistically—that is, for deciding what statistical techniques to use for a given set of variables. Some statistical methods are most appropriate for nominal-level variables, others for ordinal variables, still others for interval or ratio variables.

A *nominal variable* is measured in such a way that its values or attributes are different from one another—but just different, that's all. We cannot arrange the values in any logical or natural order; there is no underlying continuum along which scores can be arrayed. A nominal variable's values are unordered categories—like gender, for example. Males and females (the values of gender) are different from one another, but we cannot order the categories of male and female. We cannot say, for example, that men are higher or lower, better or worse, or more or less than women. There isn't any inherent or intrinsic ordering for values of a variable like gender.

Any apparent ordering of a nominal variable's categories derives only from cultural conventions. We almost always refer to the colors of the American flag as "red, white, and blue," for example, but that ordering reflects only custom or habit, not any underlying scale. After all, we could just as well describe Old Glory as "white, blue, and red" even though we almost never do so. Likewise, alphabetical orderings; they too are cultural conventions without any underlying scale.

If we assign numerals to a nominal variable's categories (e.g., designating males 1 and females 2), the numerals are arbitrary symbols. They do not imply any ranking or ordering, and we cannot do anything arithmetic with them. We cannot sensibly add, subtract, multiply, or divide them. We cannot say that two males equal one female, or that a male equals half a female. We could just as easily switch the numerals, labeling females 1 and males 2.

Gender is a *dichotomous variable*, meaning that it has exactly two values. Dichotomous nominal variables are common in the social sciences. Other examples include voting behavior distinguishing voters and non-voters, geographic location measured as rural or urban, and place of birth distinguishing native born and foreign born. But there are many nondichotomous nominal variables too such as, ethnicity, region, academic major, and favorite TV show.

So, many variables commonly used in the social sciences are nominal variables. But we social scientists also use a great many ordinal variables. An *ordinal variable* is one whose values can be rank-ordered, but that is all. The rank-ordering is based on some "natural" or "intrinsic" pattern—some underlying continuum.

Social class measured with values of lower class, working class, middle class, and upper class is an ordinal variable. There is an underlying semantic order to the values lower, working, middle, and upper. We can assign numerals to these four values indicating their positions in the ordering. For example, we might assign 1 to lower class, 2 to working, 3 to middle, and 4 to upper class. These numerals identify the relative positions of the four classes (e.g., the middle class comes between the working and upper classes just as 3 comes between 2 and 4). But we cannot perform any arithmetical operations on such numerals. We cannot say that one lower-class person plus one middle-class person equals one upper-class person. Note too that any ordered numerals would do just as well. Although there is no reason to do so, we could label lower class as 5, working class 16, middle class 43, and upper class 87. As silly as these numerals would be, they would convey the rank-ordering of these values of social class.

Unlike nominal variables, then, ordinal variables have an underlying continuum along which scores can be arranged. We can rank-order scores—that is, actual measurements—based on the ordering of values even though we cannot measure the exact differences between scores. We can say that in the social class hierarchy an upper-class person outranks a middle-class person, who in turn outranks a member of the working class, who outranks a lower-class person—a pecking order, as it were. We cannot say, however, that an upper-class person has twice as much social class as a working-class person, or that the distance between the working class and the middle class is the same as between the middle class and the upper class. We can only rank-order the values.

The ordinal nature of a variable like social class is easily recognized. Lower class "comes before" working class, which in turn "comes before" middle class, and then comes upper class. However, rank-ordering is less obvious for many dichotomous ordinal variables. Consider the civil disobedience variable described earlier—whether people should always obey the law or should on exceptional occasions follow their consciences even if it means breaking the law. With its values of either obey the law or follow one's conscience, this variable is dichotomous. But is it nominal or ordinal? At first glance, the civil disobedience variable's two responses seem simply two categories that reflect two distinct but unordered attitudes toward the law and civil disobedience. If so, then the variable is nominal. However, we can reasonably treat the obey law/follows one's conscience values as two positions on an underlying dimension running from unconditional observance of the law to unconditional compliance with one's conscience. Perhaps we could even identify intermediate positions with their own values on the variable, but we have chosen to dichotomize the variable into obey the law/follow one's conscience. Whether we regard obey the law or follow one's conscience as the lower value is surely arbitrary, but we nevertheless can regard the obey/conscience values as ordered and the civil disobedience variable as ordinal. Watch out for these dichotomous ordinal variables that at first look like nominal variables. Although some researchers do treat these kinds of dichotomous variables as nominal, often we can make better use of them by regarding them as ordinal.

In an important sense, ordinal variables convey more information than nominal variables. All variables, including nominal, distinguish differences among cases. Unlike nominal variables, however, ordinal variables also rank-order values, and in this way ordinal variables convey more information. They also allow us to array cases along a continuum. As we learn statistics, we will find that we can do more statistically with ordinal variables than with nominal variables. Ordinal variables allow us to use more powerful statistical techniques that extract more information from data. Thus we say that ordinal variables have a *higher level of measurement* than nominal variables.

Some examples of ordinal variables are attitudes like alienation measured as low, medium, or high; size measured as very small, small, average, large, and very large; and income measured as below average, average, and above average. And as we have seen, the civil disobedience variable—obey the law or follow one's conscience—is also ordinal.

An *interval variable* not only has values that can be rank-ordered, but also is measured using some fixed or standard unit of measurement such as the dollar, the year, the pound, or the inch. Take temperature measured in degrees Fahrenheit. The standard unit of measurement is the Fahrenheit degree. This fixed, constant unit of measurement allows us to perform arithmetical operations like addition and subtraction. So, for example, the difference between 70° and 75° Fahrenheit is the same as the difference between 82° and 87° Fahrenheit. Both differences are 5°. (But note that we have an *ordinal* variable if we measure temperature in broad categories such as cold, cool, lukewarm, warm, and hot. These values are rank-ordered, but with no standard unit of measurement.)

A *ratio variable* is like an interval variable in that it too is measured with a standard unit of measurement; but unlike an interval variable, a ratio variable has a nonarbitrary zero point. The zero point represents the absence of the characteristic or property being measured. Examples of ratio variables are income measured in dollars, education measured in years of schooling, and amount of television watching measured in hours per day. And speaking of temperatures, temperature measured on the Kelvin scale favored by physicists and chemists is also a ratio variable.

There is an absolute zero point for each of these variables: A person may earn no income, have no education, and watch no television, and temperature may be absolute zero on the scientific Kelvin scale. Unlike zero for interval variables like Fahrenheit temperature, the zero point for these ratio variables is not arbitrary. Zero for a ratio variable indicates absolute absence of whatever characteristic is measured—no income, no education, no television watching, and no molecular motion (in the case of 0° Kelvin). Negative values are possible for some although not all ratio variables. Someone may have a negative income, for example, indicating that the person actually *lost* more money than she made, but someone cannot have negative years of education or negative age, and temperatures cannot be negative Kelvin.

Ratio variables are called that because they allow ratios to be formed and thus permit a wider range of mathematical operations. We can say, for example, that a 20-year-old is twice as old as a 10-year-old kid and half as old as a 40-year-old person. Those sorts of ratios cannot be formed with variables measured at other levels, even interval. We cannot say that someone is twice as married as someone else, or twice as obedient to the law, or twice as intelligent.

Variables measured by counting things (e.g., dollars, years, hours, and degrees) are almost always either interval or ratio variables. The difference between interval and ratio variables is that the latter have nonarbitrary zero points.

Just as we can use more powerful statistical methods with ordinal than with nominal variables, so too do interval variables allow use of still more powerful statistics that extract even more information from data. After all, interval variables not only distinguish values (as even nominal variables do) and rank-order values (as ordinal variables do), they also describe how much difference there is between values. Interval variables thus convey more information and have an even higher level of measurement than ordinal variables.

And ratio variables convey the most information and thus have the highest level of measurement. Most basic statistics that can be used with ratio variables can also be used with interval variables, however, so we do not usually have to distinguish between interval and ratio. Moreover, there are not many truly interval variables in the social sciences. For these reasons I will combine the interval and ratio levels and refer to their variables as *interval/ratio variables*.

But we will have to distinguish nominal variables from ordinal variables and both types from interval/ratio variables. If we do not properly distinguish among these levels of measurement, we risk using statistical methods that either make too little use of data we are analyzing or make unwarranted assumptions about variables. Table 1.1 presents a summary for distinguishing among nominal, ordinal, and interval/ratio variables.

Table 1.1. Levels of Measurement of Variables

Level of Measurement	Is there an intrinsic order to values?	Is there a standard unit of measurement?
Nominal	No	No
Ordinal	Yes	No
Interval/Ratio	Yes	Yes

Distinguishing levels of measurement is not always easy. Truth be told, reasonable men and women sometimes disagree about the level of measurement of a particular variable. What may seem ordinal to one researcher may look only nominal to another. I noted this ambiguity earlier in this section with the civil disobedience variable measured by asking survey respondents if people should always obey the law or should follow their consciences even if that means breaking the law. That variable is nominal to some researchers and ordinal to others, depending on how they conceptualize civil disobedience. I am not suggesting that "anything goes" in statistics—certainly not! But statistics is not cut and dried, which is one reason it is so interesting.

Still, we can usually agree most of the time about variables' levels of measurement, and your job is to get good at making such distinctions. Let me be blunt: You *must* learn to identify levels of measurement of variables correctly in order to do statistics. Trying to do statistics without distinguishing among levels of measurement is like trying to cook without distinguishing between fruits and veggies. You might be able to concoct something, but it won't taste very good. Imagine using potatoes instead of lemons in a meringue pie, or cherries instead of beans in Tex-Mex tacos. So work very hard on identifying the levels of measurement of variables.

You will find that correctly identifying levels of measurement becomes much easier with practice. The workbook accompanying this text offers many variables to practice with. Levels of measurement many seem a little confusing now, but you will quickly become good at recognizing each level. Doing so is easier if you keep in mind that level of measurement refers not to a variable in the abstract, but rather to a variable measured in a particular way. Income, for example, can be measured as either an ordinal variable (e.g., low, middle, high income) or an interval/ratio variable (e.g., dollars per year). To determine level of measurement, then, we need to know not only how a variable is conceptualized but also how it is actually measured. The key to determining level of measurement is inspecting the variable's values. Are they inherently ordered? If not, the variable is nominal; if values are ordered, then the variable is either ordinal or interval/ratio. Is there a standard unit of measurement? If so, the variable is interval/ratio.

1.6   Mutually Exclusive and Collectively Exhaustive

Researchers usually find variables most useful if their values are both mutually exclusive and collectively exhaustive. *Mutually exclusive* means that the values do not overlap—each case has only one value. For example, a person can be Protestant *or* Catholic, but not *both* Protestant *and* Catholic. Likewise, gender distinguishing male and female has mutually exclusive values. Anomalies aside, a person cannot be both male and female. (Of course, someone can be both Protestant and male. No problem. Mutually exclusive is a quality that applies only to values of the same variable.)

We need to be careful that variables at all levels of measurement have values that are mutually exclusive, but ordinal and interval/ratio variables offer special temptations to violate this principle. It is tempting, for example, to classify ages as, say, 0–10, 10–20, 20–30, and so on. But note that a 10-year-old could be in two categories. Likewise a 20-year-old. The values overlap and thus are not mutually exclusive. It's much better to use mutually exclusive values such as 0–9, 10–19, 20–29, etc.

Collectively exhaustive means that the set of values includes all cases—every case falls into some category. For instance, religious categories such as Protestant, Catholic, Jewish, none, and other include all possible answers to a question about religious preference. The residual value "other" ensures that every case falls into a category since people who are not Protestant, Catholic, Jewish, or without religious affiliation are bound to be some "other" religion.

Thus, if a variable's values are both mutually exclusive and collectively exhaustive, every case can be put into one and only one category. Religion measured as Protestant, Catholic, Jewish, none, or other is both mutually exclusive and collectively exhaustive. So is sex measured as male or female. Or social class measured as lower, working, middle, or upper class. Or civil disobedience measured as obey the law or follow one's conscience.

1.7   Continuous and Discrete Variables

While identifying levels of measurement of variables, we sometimes find it useful to distinguish also between continuous and discrete variables. A *continuous variable* may in principle take on any value in its range of possible values. For example, with age measured in small enough units, a person passes through every possible age from birth to death. Sure, in Western cultures we usually round age down to the last integer. We say someone is 19 years old. That's precise enough for most purposes. But in principle we could describe someone's age as 19 years, 3 months, 7 days, 16 hours, 23 minutes, 8 seconds . . . right down to the micro-nanosecond and beyond.[3] Yes, age measured in standard units like hours, seconds, and fractions thereof is a continuous variable. Other examples of continuous variables are most attitudinal variables such as prejudice, liberalism, and alienation. They too have underlying continua that, in principle, can be broken down into a set of measurements as fine or precise as we want.

---

[3] I distracted myself in boring junior high classes by calculating my age to the exact second. (Maybe you did the same thing.) Of course, my very precision ensured that my calculated age was incorrect the instant I determined it.

A *discrete variable*, on the other hand, can have only certain values within its range. Family size is an example. A family may have 1 member or 2 members or 3 members and so on, but cannot have, say, 1.7 members. Family size can take on only integer values 1, 2, 3 . . . and cannot have any values between these integers. (Please, let's not confuse this example with pregnancies or half-brothers or half-sisters.) Other examples of discrete variables are number of CDs owned, number of college courses taken, and number of friends of a different race.

Nominal-level variables are always discrete, but ordinal and interval/ratio variables can be either discrete or continuous. The distinction is sometimes tricky to draw because continuous variables may *look* discrete since we must necessarily measure such variables using rounded values. The General Social Survey measures outlook on life by asking respondents if they find life exciting, pretty routine, or dull. Just three values. But in principle people can have any value along a continuum of outlook on life, so outlook on life is really a continuous variable. The continuous-discrete distinction is also a little tricky since it is often useful to treat a discrete variable that has a large number of values—income, for example, or a country's population— as if it were continuous even though, strictly speaking, it is discrete.

But however difficult to make, the continuous-discrete distinction matters. Just as with levels of measurement, we will find that some statistical techniques are appropriate for continuous variables and others for discrete variables. We need to identify variables as continuous or discrete in order to choose the right statistical procedures for the particular data we are analyzing.

1.8   What Cases, Variables, and Data Files Look Like

I want you to see what cases and variables in a data file actually look like when we analyze them. The U.S. *General Social Survey*, a national survey of adult Americans, offers a good example. Social scientists use this social survey so much that we usually abridge its name to just *GSS*. The GSS is conducted by the National Opinion Research Center (NORC) using rigorous, sophisticated random sampling procedures to select its respondents. NORC also thoroughly trains and carefully monitors its interviewers. The GSS has been conducted almost every year since 1972. National survey data don't get any better than this.[4] Even as you read this sentence, social scientists

---

[4] Canada has its own General Social Survey. In fact, at last count, some 29 countries have national surveys comparable to the U.S. and Canadian General Social Surveys. These high-quality surveys coordinate their questionnaires to facilitate comparative research.

from Cambridge to Palo Alto (and overseas too) are analyzing exactly the same data that we will use throughout this text. Talk about equality of access—that's what the GSS gives us.

The particular variables and the number of cases in the GSS vary from year to year, but let me show you what the information in the 1996 GSS data file looks like. This GSS has 1046 variables and 2904 cases. It is conventional to arrange data files so that rows correspond to cases and variables are in columns. Table 1.2 presents the case numbers plus coded scores on 9 variables for the first 10 GSS cases. The three-dot ellipses indicate that only a part of the data set is shown. A full display of the GSS data set would be 1046 variables wide and 2904 cases long—far too much to show on a page.

Table 1.2. First 9 Variables for First 10 Cases in the General Social Survey

Case Number	General Social Survey Variable									
	WORK-ING	PRES-TIGE	MARI-TAL	DI-VORCE	DAD PRES	MA PRES	# SIBS	# CHIL-DREN	AGE	...
0001	5	32	1	2	28	−9999	2	0	79	...
0002	1	32	5	−9999	44	64	1	0	32	...
0003	4	30	3	−9999	−9999	−9999	2	3	55	...
0004	2	46	1	1	−9999	−9999	98	4	50	...
0005	7	28	4	−9999	39	−9999	7	4	56	...
0006	1	52	4	−9999	22	−9999	4	4	51	...
0007	1	32	4	−9999	−9999	23	1	3	48	...
0008	1	34	5	−9999	−9999	−9999	4	1	29	...
0009	1	34	4	−9999	−9999	17	6	0	40	...
0010	7	−9999	4	−9999	23	−9999	3	4	46	...
.	.	.	.	.	.	.	.	.	.	...
.	.	.	.	.	.	.	.	.	.	...

Names of variables head each column. You can probably guess what most variables are just from their names. WORKING is the respondent's employment status—full time, part time, unemployed, etc. PRESTIGE is the respondent's occupational prestige, MARITAL is the respondent's marital status, and so on. You can see that all information has been converted into numbers called *codes*. Computers work more efficiently with numeric than with alphabetic information. The codes for some variables are obvious. For # SIBS and # CHILDREN, codes are obviously numbers of siblings and children, although that 98 for # SIBS looks odd (we will learn later that it is the code for Don't Know). AGE certainly reports respondent's age in years. Codes for some variables are not so obvious. For example, for the variable MARITAL, married respondents are coded 1, widowed respondents are coded 2, etc.—but you would not know that from just looking at the codes. Later on, we will learn how to find out what all the code numbers stand for.

We can read across rows to find a given case's scores on variables. For example, you can see that Case 0001 is a 79-year-old with 2 siblings and no children. If you knew the codes for the other variables, you would also know that case 0001 is retired (no surprise for this near-octogenarian), is married, has never been divorced, and had a job with below average prestige.

What about the –9999 code? This is a special code meaning that the variable does not apply to the respondent. For example, –9999 for DIVORCE (based on the question "Have you ever been divorced or separated?") may mean that the respondent has never married or is now divorced or separated. Either way, it makes no sense for the GSS interviewer to ask the respondent if he/she has ever been divorced —the question is inapplicable. A –9999 code may also mean that a given survey question was simply not asked for the respondent even though it might apply. The GSS asks some questions of all 2904 respondents. For example, most demographic questions like age, education, and marital status, are asked of everyone. However, many other interview questions concerning beliefs and attitudes are asked on a rotating basis so that the question is asked only of a subsample of respondents. For example, less than half of GSS respondents—1332 cases—were asked if they thought people should always obey the law or should follow their own consciences, and somewhat less—1252—gave usable responses. (The other 80 respondents said they couldn't choose between obey the law and follow one's conscience or else gave no answer at all to the question.) Rotating questions this way allows the GSS to include far more questions and still have enough responses for reliable statistical analyses.

Far less common, codes of 98 or 99 may mean that the respondent didn't know the answer or refused to answer a question. Or perhaps, very rarely, the interviewer may have inadvertently forgotten to ask the question or record the answer. Missing data of these kinds (especially –9999 for inapplicable or rotating variables) are fairly common in surveys and other data sources used by social scientists. We'll learn throughout this text and workbook what to do about missing data. Mostly we will exclude missing data from our analyses, and so we will necessarily find statistical analyses based on varying numbers of cases even though the analyses use the same data set.

1.9   Aggregate Data

Cases in the General Social Survey are individual people, and that is often the sort of data social scientists work with. Sometimes, however, it is advantageous to combine individual scores into larger groupings. For example, we may use information about individual students at colleges to report the percentage female at each college; our cases then are colleges, not individual students. Or data from individual respondents in a national survey may be combined into average responses for each state, thereby creating a data set in which cases are the 50 states. Similarly, cross-cultural researchers may use data sets in which cases are the 50 most populous countries rather than individual people. Cases might be Algeria, Argentina, Australia through Venezuela, Vietnam, and Zaire. Data of this kind in which cases are larger units of analysis are called *aggregate data*.

If the larger units are spatial or geographic areas like states, provinces, or countries, the aggregate data are *ecological data* and their variables are *ecological variables*. Some examples of ecological variables are murder rate, percent urban, average income, and percent Hispanic in the 50 states. Because they pertain to geographic units, ecological variables have the advantage that we can display them visually on maps. So, for example, we can create a map of the United States with each state color-coded to represent, say, its murder rate. You have probably seen such maps in publications like *Time* or *USA Today*. They help us visualize how a variable is distributed geographically. We might notice, for example, that higher murder rates tend to cluster in certain regions of the United States while other regions have low rates. We will learn how to produce these kinds of maps as we learn statistics.

Aggregate data, especially ecological data, are very useful, even more useful for some purposes than data from smaller units of analysis. Aggregate units such as states, provinces, or countries often are inherently or substantively interesting. Understanding how a variable such as unemployment is distributed across states may, for example, guide economic policies and planning. Furthermore, many social scientific theories apply to aggregate units rather than individuals. Durkheim's theory of suicide, for example, or Michels' Iron Law of Oligarchy ("who says organization says oligarchy") apply to aggregate units. Social scientists are often interested in averages such as average income and rates such as homicides per 100,000 population, and averages and rates by definition pertain to aggregate units. Moreover, variables for relatively rare, low-incident characteristics or behaviors may be almost impossible to detect or "pick up" at the individual level. Such a small proportion of the population are psychiatrists that even a large survey like the General Social Survey will include very few if any psychiatrists, but at the aggregate level we can consider the number of psychiatrists per 100,000 population in each state—an aggregate variable.

We must always be careful not to confuse individual and aggregate levels of analysis. In particular, we must guard against inferring individual characteristics from analysis of aggregate data or, for that matter, inferring aggregate characteristics from data for individuals. Each entails a logical error. The logical error of inferring characteristics of individuals from an analysis of aggregate data is known as an *ecological fallacy*. Take an example: states with larger proportions of Jewish residents have higher rates of illiteracy. However, it would be a mistake—an ecological fallacy—to infer that Jews are likely to be illiterate. Indeed, Jews have a very *low* illiteracy rate. It is just that Jews tend to live in states with higher proportions of illiterates even though very few Jews are themselves illiterate.

Aggregate data sets often have fewer cases than data sets based on individuals. Whereas surveys typically have hundreds or even thousands of cases, data for American states have 50 cases. Data for Canadian provinces and territories have only 12 cases. Aggregate data based on districts of cities usually have fewer than 100 cases. Data sets for countries usually have no more than about 160 cases, and often fewer. As I noted earlier, we will use two aggregate data sets in this text: one for the 50 American states and another for the 50 most populous countries. While fewer cases prevent us from using some statistical techniques, they allow us to make even greater use of still other techniques, particularly those that display data visually on maps. As in the rest of life, in statistics sometimes less is more.

Aggregate variables are more likely to be continuous and interval/ratio than are variables using individuals as a unit of analysis. Many variables for individuals are nominal or ordinal—their sex, for example, or race, or religion. Aggregate variables are more likely to be rates, averages, or percentages—murders per 100,000 population, for example, or average income, or percent living in poverty. Variables like these are continuous interval/ratio variables, and thus allow us to use statistical procedures not possible with discrete nominal or ordinal variables.

1.10   MicroCase

As this text's subtitle promises, we will learn statistics using *Student MicroCase*. Student MicroCase is a computer program that makes it fast and easy to do statistics even with very large data files that have hundreds of variables and thousands of cases. MicroCase is a particularly well-designed piece of software. Although it hasn't the range of statistical techniques offered by some other well-known statistical packages like SPSS-X, SAS, or BMD, MicroCase can do just about any sort of statistical analysis that you are ever likely to want to carry out. Moreover, MicroCase is friendlier than these other packages, and it is mighty quick too.

You will like MicroCase a lot. Here are some of the things that MicroCase can do:

- Compute percentages.
- Produce statistical tables.
- Produce pie charts and bar graphs.
- Draw outline maps showing geographic distributions of variables.
- Compute means and medians.
- Compute standard deviations and variances.
- Compute cross-tabulations, chi squares, and measures of association.
- Do t tests and analyses of variance.
- Compute regression and correlation coefficients.
- Compute multiple regression and correlation coefficients.
- Create new variables from old ones and manage data in useful ways.

There's no need to worry if you don't know what some of these things are. You will know all of them after finishing this text and accompanying workbook.

MicroCase is available for DOS and Windows compatible computers like IBM, Dell, and Compaq computers. (Sorry, fellow Apple fans, but we have to use DOS or Windows computers for MicroCase.) You already own MicroCase. A version called Student MicroCase came on a diskette with this text's workbook. Student MicroCase is a slimmed down but still powerful version of the full MicroCase.[5]

Don't worry if you don't know much about computers or even if you have never used one. You don't need to know much about computers themselves to make good, intelligent use of MicroCase to do statistics. You don't have to write computer programs or design computer systems—nothing like that. The workbook will give you all the information you need.

---

[5] Your school may have the full MicroCase if it participates in the MicroCase Curriculum Plan or has a site license. You can find out from your instructor if the full MicroCase is available at your school.

1.11   Ideas and Thinking

Statistics is extremely important in research, but research (unlike this sentence) neither begins nor ends with statistics. In the social research process, statistics comes somewhere in the middle. *Ideas* come first and last because what good researchers care about are ideas. Ideas about the world and how it works—that is what initially and ultimately matters.

And where do ideas come from? Lots of places. Ideas may spring from simple, even unsystematic observations of the world. Ever notice how women tend to sit near classroom doors and men tend to sit farther inside a classroom? Hmmm . . . an observation like that might lead to some ideas about gender roles.

Or ideas may derive from theories. So, for example, feminist theory may prompt some ideas about child-rearing practices, or symbolic interactionism may lead to ideas about roles and labeling, or world systems theory may generate ideas about nations' economic situations. Lots of theories grow in the social scientific garden, each blooming with ideas.

Ideas may come from what we read. Plato's *The Crito* or Thoreau's essay *Civil Disobedience* or Martin Luther King's *Letter from the Birmingham Jail* may start us thinking about civil disobedience and what sort of people are most likely and least likely to follow their own consciences even if it means breaking the law.

Or ideas may grow from the need to solve nitty-gritty, practical problems. Concerned about improving race relations on campus? Thinking about what might be done may generate ideas about intergroup relations. Troubled about pollution of local streams? Thinking about solutions may generate ideas about support for environmental initiatives. Disturbed by students' political apathy? Thinking about programs to overcome apathy may generate ideas about its social sources.

Wherever ideas come from, they shape the research process, including statistical analysis. Ideas suggest what questions to pose, what issues to address, what hypotheses to test. Ideas identify what cases should be studied. Ideas describe what variables should be used and how variables should be measured. Ideas suggest how variables may be related to one another. And, crucial for our purposes, ideas direct us toward appropriate statistical techniques.

So ideas come first and affect the research process, including statistical analysis. But statistics in turn affects ideas. Statistical analyses lead researchers to investigate new issues, develop new ideas, generate new hypotheses, seek new cases to study, introduce new variables, and explore different ways to interrelate variables. What we have, then, is not a research circle that ends where it started, but rather a research spiral, with ideas leading to research and statistical analyses, which in turn lead to new ideas that lead to additional research and analysis, and on and on. We do not expect to arrive at the "truth," but we certainly intend to get closer to it.

Yes, ideas matter. And ideas, of course, require *thinking*. We need to do a lot of thinking in statistics. To be sure, statistics of a crude sort can be done with little thought by just mechanically applying statistical procedures. But the result would be a dumb analysis that does not tell us much about the world and how it works. To do statistics right requires thinking—thinking hard. So be ready to think a lot any time you do statistics.

1.12   Playing with Data

To do statistics well, you need to *play* as well as think. Fool around with data the way you played with fingerpaints when you were a kid. Remember how you fingerpainted weird flowers using just your index finger? Then you wondered what would happen if you used two fingers, so you smeared out the flowers and created new, even weirder flowers mixing in new colors. And then you wondered what would happen if you used your whole palm, and found those flowers weren't so interesting, so you smeared them out and drew a goofy looking face using just your little finger. And then you . . . .

You played with fingerpaints, and that's what you should do with data. Try things out. See what happens. Follow your curiosity. Try changing the analysis a little . . . or a lot. Look for patterns in the data. See what happens if you bring another variable in. Try different variables. Moving data around with the computer is just about as easy as moving fingerpaints around with your fingers.

But I don't mean to push the fingerpainting-statistics analogy too far. Fingerpainting lacks the discipline and intellectual rigor that characterizes genuine art. Fingerpainting is, in the end, only self-indulgent expression. That's why fingerpaintings hang on family refrigerators rather than museum walls. You need to supplement your playfulness with much discipline and intellectual rigor to carry out statistical analyses effectively. That's where the thinking described in the previous section comes in. But the playful quality is essential and complements discipline and rigor. As in the rest of your life, you will learn and do statistics better and your data analyses will be more fruitful if your approach is playful.

1.13   What to Expect in This Text

Let me explain how I have organized this text. We will learn statistics by beginning with the simplest situations and then moving through increasingly complex analyses. We will start with methods for analyzing only one variable at a time. Then we will take up methods for analyzing a relationship between two variables. That's a little more complex. And finally we will consider techniques for studying relationships among three or more variables at a time. That's really complex. Sensibly enough (at least if you know some Latin), these three situations involve *univariate, bivariate,* and *multivariate analyses,* respectively. That is how we will proceed in this text. The accompanying workbook has exercises that encourage you to apply your knowledge of statistics and carry out computer analyses of real data.

This book covers basic statistical methods. Although you will not be a full-fledged statistician after finishing *Social Statistics,* you will know enough statistics to critically evaluate others' uses of basic statistical techniques and to analyze data yourself. In other words, you will be both an informed consumer and a responsible producer of statistics. You will also be well prepared to study more advanced statistics should you choose to do so. In short, you won't learn everything about statistics from this text, but you will learn a lot.

I have written this text for students in the social sciences and related disciplines such as social work, criminal justice, public administration, and education. I assume that you need to carry out real analyses using real data, so I do not avoid the problems you and I find in the "real world"—missing information, skewed data, outliers, variables that need to be collapsed, and so on. In this text we will face the messiness that makes the social world so challenging to study and so much fun to live in. This text relies heavily on real data from sources like the U.S. Census, *Uniform Crime Reports,* and the General Social Survey—data that full-fledged social scientists analyze in their own research. Thus, we can't avoid facing the real issues and problems that come with real data. Still, there are times when messiness of data can get in the way of learning statistics, so I have not hesitated to make up data to introduce certain statistical techniques when "clean" examples foster learning. We'll use whatever data—real or imaginary—best helps us learn statistics.

I also assume that you need to present your statistical analyses to others (your instructor now, employers or a larger "public" later in your life), so I include guidelines for writing up statistical analyses, formatting tables, and constructing graphs. You will also find examples of analysis write-ups following several chapters. You will probably find these write-ups useful as models for describing your own research results. You don't need to be a professional statistician to communicate statistical findings effectively to others. You will be able to do so after finishing this text.

So this text does not skimp on nitty-gritty details about doing real statistics in the real world. But I am also concerned that you understand statistical principles and reasoning. You need this understanding to know what statistical procedures are right for given situations and to interpret statistical analyses sensibly. Therefore, I will go beyond formulas and procedures to explain why we are doing what we do. You will find almost no computational formulas in this book to facilitate calculations. Computers render such formulas unnecessary. There is, of course, a place for computational formulas. They mostly belong in a Museum of Statistical Antiquities, maybe in a glass case next to slide rules, to remind us how unpleasant and tedious statistics used to be.

But this text does not shy away entirely from formulas and calculations. To the contrary, we will learn to do calculations "by hand" before using the friendly computer to do this mostly routine work for us. Initially while learning data analysis, calculating statistics with our human brains is essential for understanding why statistics work the way they do and what the computer's faster brain does for us. However, we will use definitional rather than computational formulas. Although less convenient and more tedious, definitional formulas describe what statistics are really about and thus are the best way to understand how statistics works. The *Doing Statistics Using MicroCase* workbook accompanying this text invites you to do paper-and-pencil calculations before carrying out computer analyses.

But in the last analysis (literally), the computer exercises in the *Doing Statistics Using MicroCase* workbook are even more important than the "by hand" exercises. I have written the workbook for you to use actively with a computer. You should, in fact, read and do most of the workbook while using a computer.

You will find this textbook and workbook conversational in style. That's deliberate. It's the way I teach. My students and I interact in the classroom and, within the limits imposed by print, there is no reason why you and I should not interact via this text. So, I hope you will not take my use of the second-person "you" as an unwarranted familiarity, nor my frequent use of "we" as either an affectation or a ploy. You and I—*we*—are learning statistics together even though you are learning stat for the first time and I am learning it for the umpteenth time.

When you are finished with this book, I hope you will begin a lifetime using statistics effectively. I trust that you will be an intelligent critic of others' applications of statistics and will use statistics intelligently yourself. But beyond these practical goals, I hope you will appreciate the beauty, elegance, and grace of statistics. Like other magnificent achievements of human reason, statistics at its core is an aesthetic endeavor.

1.14   Summing Up Chapter 1

Here is what we have learned in this chapter:

- The word *statistics* refers both to numerical summaries of information and to methods used to summarize and generalize information.

- Data are records of observations.

- Usually we organize data into data sets and data files to analyze using a computer.

- A unit of analysis is the person, object, or event that a researcher is studying. Cases are the actual units from which data are collected.

- There are two branches of statistics: descriptive statistics, which summarizes information; and inferential statistics, which generalizes from sample data to the larger population and allows us to assess our confidence in our findings.

- We often rely on sample data because it is often infeasible or impossible to use data from an entire population. Sample data are often more complete or of higher quality than population data.

- A parameter is a characteristic of a population. A statistic is a characteristic of a sample.

- Variables are characteristics or properties that take on one or more values.

- Measurement is the process of finding the values of a variable for different cases. These measured values are called scores.

- A scale is a set of values of a variable.

- We decide what statistical techniques to use in part by the level of measurement of variables: nominal, ordinal, or interval/ratio.

- Nominal variables have unranked categorical values.

- Ordinal variables have values that can be rank-ordered but are not based on a standard unit of measurement.

- Interval/ratio variables have rank-ordered values based on a standard unit of measurement.

- The level of measurement of a variable depends on how a variable is actually measured.
- Knowing the level of measurement of variables is essential for choosing appropriate statistical techniques.
- A continuous variable can, in principle, take any value within its range. A discrete variable can take only a limited number of values.
- Some statistical techniques are more appropriate for continuous variables, others for discrete variables.
- Aggregate data are based on units of analysis larger than individuals.
- Aggregate variables are more likely to be interval/ratio and to allow a wider range of visual displays of data.
- Ecological data are based on spatial or geographic units of analysis and can be displayed on maps.
- We need to guard against the ecological fallacy of inferring individual characteristics from aggregate data.
- Thinking matters (so do a lot of it).
- The best way to approach data analysis is with a playful attitude combined with intellectual rigor.

Key Concepts and Procedures

Ideas and Terms

statistics (as information)
statistics (as methods)
data
data set
data file
units of analysis
case
descriptive statistics
inferential statistics
sample data
population data
census
parameter
statistic
variable
measurement
scale
score

levels of measurement
nominal variable
dichotomous variable
ordinal variable
interval/ratio variable
mutually exclusive
collectively exhaustive
continuous variable
discrete variable
General Social Survey (GSS)
codes
aggregate data and variables
ecological data and variables
ecological fallacy
MicroCase
univariate statistics (univariate analysis)
bivariate statistics (bivariate analysis)
multivariate statistics (multivariate analysis)

CHAPTER 2
Frequency and Percentage Distributions

In this chapter I will describe how to summarize information about variables taken one at a time. In other words, we will deal with univariate situations. Here we will learn about frequency and percentage distributions as well as pie charts, bar graphs, and map displays. The accompanying chapter in the workbook describes how to use MicroCase to carry out these univariate statistical analyses and offers you practice exercises.

After this chapter you will be able to:

1. Transform raw data into univariate frequency distributions.

2. Transform univariate frequency distributions into percentage distributions.

3. Interpret univariate percentage distributions.

4. Understand that percentages are standardized frequencies.

5. Create and explain cumulative percentage distributions.

6. Create and interpret presentation-quality univariate percentage tables.

7. Apply general guidelines for collapsing the values of variables.

8. Recognize missing data to exclude from analyses.

9. Recognize that analyses may be based on subsets of cases.

10. Create presentation-quality pie charts and bar graphs.

11. Explain the effects of levels of measurement on the use of cumulative distributions and the choice of pie charts or bar graphs.

12. Explain what an outlier is and recognize outliers in distributions.

13. Create and interpret maps of ecological variables.

2.1   Frequency Distributions

An easy, straightforward way to summarize information about a variable is to count the number of cases with each score. This summary of the pattern of variation of a variable is a *frequency distribution*. Instead of thinking about each individual score, we need to think only about how many cases have each score.

Let's begin with a couple of simple examples. Consider the variable civil disobedience introduced in Chapter 1. I noted that the General Social Survey asked respondents if people should obey the law without exception or whether there are exceptional occasions on which people should follow their consciences even if it means breaking the law. For convenience I will refer to this variable with the shorthand name Civil Disobedience. Table 2.1 reports scores on Civil Disobedience of 50 actual GSS respondents.

Table 2.1. Scores on Civil Disobedience (50 Cases)

Case Number	Civil Disobedience	Case Number	Civil Disobedience
01	Follow Conscience	26	Follow Conscience
02	Follow Conscience	27	Follow Conscience
03	Obey Law	28	Follow Conscience
04	Obey Law	29	Obey Law
05	Follow Conscience	30	Obey Law
06	Obey Law	31	Obey Law
07	Obey Law	32	Follow Conscience
08	Follow Conscience	33	Obey Law
09	Follow Conscience	34	Obey Law
10	Follow Conscience	35	Follow Conscience
11	Follow Conscience	36	Follow Conscience
12	Obey Law	37	Follow Conscience
13	Obey Law	38	Obey Law
14	Obey Law	39	Obey Law
15	Obey Law	40	Follow Conscience
16	Follow Conscience	41	Follow Conscience
17	Follow Conscience	42	Follow Conscience
18	Follow Conscience	43	Follow Conscience
19	Obey Law	44	Follow Conscience
20	Obey Law	45	Obey Law
21	Obey Law	46	Follow Conscience
22	Follow Conscience	47	Obey Law
23	Follow Conscience	48	Follow Conscience
24	Follow Conscience	49	Obey Law
25	Follow Conscience	50	Follow Conscience

These 50 cases are a sample of the GSS sample, as it were. It will be much easier to learn about frequency and percentage distributions if we work with just these 50 cases rather than all 2904 cases in the GSS. The first two cases believe in following one's conscience, the third and fourth cases believe people should always obey the law, the fifth case believes people should follow their conscience, and so on. These are *raw data*—the scores that we begin with.

We can tally—that is, count up—the number of cases that have each score, as in Table 2.2. This tally is a frequency distribution showing how many respondents answered obey the law and how many answered follow one's conscience. The lower-case f heading the right-hand column is the conventional abbreviation for *frequency* in statistics. Here I have reduced 50 pieces of information to two numbers—counts for each of the two values of the Civil Disobedience variable. We see readily that 21 respondents believe in always following the law and 29 respondents believe there are occasions on which people should follow their consciences even if it entails breaking the law. Thus, somewhat more people endorse than reject civil disobedience.

Table 2.2. Civil Disobedience Tally

Civil Disobedience	Tally	f
Follow Conscience	/ / / / / / / / / / / / / / / / / / / / / / / / / / / / /	29
Obey Law	/ / / / / / / / / / / / / / / / / / / / /	21
Total		50

We need to be able to communicate information clearly and succinctly, and there is no sense troubling others with our tally marks. Table 2.3 takes the tally and turns it into a good-looking, presentable *frequency table*.

Table 2.3. Civil Disobedience
(in frequencies)

Civil Disobedience	f
Follow Conscience	29
Obey Law	21
Total	50

Now consider a second example. Suppose these same 50 respondents were also asked how many years of education they completed. Table 2.4 presents responses for the same 50 GSS cases categorized into five broad educational levels.

Table 2.4. Scores on Education (50 Cases)

Case Number	Education	Case Number	Education
01	High School	26	Graduate Degree
02	B.A.	27	< High School
03	< High School	28	B.A.
04	High School	29	< High School
05	High School	30	High School
06	High School	31	< High School
07	B.A.	32	High School
08	High School	33	High School
09	High School	34	Junior College
10	High School	35	High School
11	Graduate Degree	36	B.A.
12	High School	37	Graduate Degree
13	High School	38	High School
14	High School	39	< High School
15	< High School	40	High School
16	High School	41	B.A.
17	High School	42	High School
18	Junior College	43	High School
19	Graduate Degree	44	< High School
20	< High School	45	High School
21	High School	46	< High School
22	B.A.	47	B.A.
23	Junior College	48	B.A.
24	High School	49	High School
25	B.A.	50	< High School

Table 2.5 reports frequencies. I have left out the tally marks and given the table a title, just as we would if we were presenting this information to others. This frequency table reduces 50 bits of information to only five numbers (not counting the total). We find that 4 respondents have graduate degrees, 9 graduated from college, 3 completed junior college, and so on.

Table 2.5. Education Completed
(in frequencies)

Education	f
Graduate Degree	4
B.A.	9
Junior College	3
High School	24
Less Than High School	10
Total	50

Tallying cases to produce a frequency distribution is as easy as . . . well, as easy as one, two, three. All you do is count. Sure we have lost some detail found in the data we began with since we no longer know the score of each individual case. But that sort of detail gets in the way of describing the distribution of the variable, and the frequency distribution helps us make far better sense of our data. We now know how many cases have each score and thus understand how scores are distributed.

2.2 Percentage Distributions

Frequency distributions are helpful in summarizing information, but sometimes they are difficult to interpret. A frequency distribution for a large number of cases presents large frequencies, and those large frequencies may be difficult to grasp. We can handle small frequencies like 4, 9, and 3 easily enough, but what if frequencies are in the hundreds or thousands or even more? Most of us find it hard to think in terms of very large numbers.

Moreover, comparison of two or more frequency distributions is difficult if the distributions are based on different numbers of cases. For example, it would be hard to compare the educational frequency distributions of Texas and Vermont since they have such different population sizes. The U.S. Census reports that Texas has 2,204,099 people with college degrees compared with Vermont's 91,522, but what can we conclude about the relative educational achievements of

Texans and Vermonters? Texas has far more college grads, but, after all, Texas also has a much larger adult population than Vermont—16,986,335 compared with only 562,758. Are Texans more likely or less likely to have college degrees? It is hard to say from just frequencies.

We can handle these problems if we standardize our summary distribution by calculating what each frequency would be *if* there were a total of exactly 100 cases. This standardization is called *percentaging*. We are so used to percentages that we rarely think about what they actually are: Percentages are what frequencies would be *if* there were a total of 100 cases. Percentaging reduces the magnitudes of large frequencies to manageable numbers (i.e., percentages) that range from 0 to 100. By standardizing each frequency on the same base—100—we can easily compare percentages. We can determine, for example, that Vermont has relatively more college graduates than Texas: 16 percent of adult Vermonters compared with 13 percent of adult Texans graduated from college.[1]

You learned to percentage in elementary school, but to refresh your memory, here's how to do it:

1. Divide each frequency by the total number of cases.

2. Multiply that result by 100.

As a formula:

$$\text{Percent} = \frac{f}{N}(100)$$

where f = frequency

N = total number of cases

For example, of the 50 cases described in the previous section, 29 cases, or $\frac{f}{N}(100) = \frac{29}{50}(100) = 58$ percent, endorse following one's conscience even if it means breaking the law. Similar arithmetical razzle-dazzle reveals that 21 cases, or $\frac{21}{50}(100) = 42$ percent, believe in always obeying the law. A set of percentages always sums to 100 (unless rounding error makes the sum slightly lower or higher.)

If we only divide a frequency by N (i.e., don't multiply by 100), we have a *proportion*. So, the proportion of cases that responded follow one's conscience is .58 and the proportion that responded obey the law is .42. Just as percentages vary between 0 and 100, proportions

---

[1] *Per cent*, from the Latin *per centum* meaning "by the hundred," can be spelled as either one word or two. One word has become more common, however, and I will use it in this text.

vary between 0 and 1.00. And just as a set of percentages always sums to 100, a set of proportions always sums to 1.00 (again, assuming no rounding error). Percentages are, of course, simply proportions multiplied by 100. Percentages and proportions present the same information—they just do so using different bases (100 and 1, respectively). Conventions and preferences for either percentages or proportions vary across disciplines and even subdisciplines of researchers. Mostly, though, social scientists rely on percentages, and that's what we'll do in this text.

Table 2.6 is a *percentage table* for the distribution of our 50 responses to the civil disobedience question. We find that a higher percentage of respondents approve of following one's conscience than always obeying the law—58 to 42 percent.

Table 2.6. Civil Disobedience
(in percentages)

Civil Disobedience	Percent
Follow Conscience	58
Obey Law	42
Total	100
(N)	(50)

The parenthetical (N) at the bottom reports the total number of cases on which percentages are based.[2] In statistics an upper-case N usually stands for the total number of cases.

If we percentage the frequency distribution for education, we get Table 2.7. This table reports that almost one-fifth (18 percent) of survey respondents are college graduates.

Table 2.7. Education Completed
(in percentages)

Education	Percent
Graduate Degree	8
B.A.	18
Junior College	6
High School	48
Less than High School	20
Total	100
(N)	(50)

---

[2] Sometimes a lowercase n is used for number of cases, especially for sample rather than population data. However, I will use an uppercase N throughout this text.

Almost half (48 percent) are high school graduates who did not attend college. A fifth (20 percent) have less than a high school education.

A word of caution: Be careful about percentaging if the total number of cases is small. Percentages are unstable and unreliable with a small N, so we can't trust percentages if the N is small. Movement of one or two scores from one value to another will produce large changes in percentages. With an N of 50 cases, each case accounts for 2 percentage points. In Tables 2.6 and 2.7, therefore, the shift of a single score from one value to another would lower one percentage 2 points while raising another 2 points. Statisticians differ in their definitions of a "small" N. Some caution against percentaging when the total number of cases is less than 30 or so. Others regard an N of 50 or even 100 as a minimum for reliable percentages. All, however, advise against trusting percentages based on few cases. (Obviously I am skirting dangerously close to and may have even crossed over the edge of these guidelines in order to keep examples simple. I promise not to do so very often.)

2.3   Cumulative Distributions

A *cumulative percentage* is the percentage of all scores that have a given value or less. To find a cumulative percentage:

1. Add all frequencies for the given value and all lesser values.

2. Divide that sum by the total number of cases.

3. Multiply that result by 100.

As a formula:

$$\text{Cumulative percent} \ = \ \frac{F}{N}\,(100)$$

where F = cumulative frequency (i.e., the sum of frequencies of a given value and all lesser values)

N = total number of cases

Equivalently (although risking more rounding error), we can find a cumulative percentage by adding up the individual percentages for a given value and all lesser values. Although statisticians use an upper-case F in several quite different ways, here it stands for the cumulative frequency, or the sum of all frequencies of a given or lesser value.

Table 2.8 is the cumulative percentage table for our hypothetical education data. For example, with frequencies taken from Table 2.5, the cumulative percentage of cases with a high school education or less is given by

$$\text{Cumulative percent} = \frac{F}{N}(100)$$

$$= \frac{10 + 24}{50}(100)$$

$$= \frac{34}{50}(100)$$

$$= 68$$

And for junior college or less:

$$\text{Cumulative percent} = \frac{10 + 24 + 3}{50}(100)$$

$$= \frac{37}{50}(100)$$

$$= 74$$

Table 2.8. Education Completed
(in cumulative percentages)

Education	Cumulative Percent
Graduate Degree	100
B.A.	92
Junior College	74
High School	68
Less than High School	20
(N)	(50)

We can also find this cumulative percentage for junior college or less by adding 20 (the percentage with less than high school) and 48 (the high school percentage) and 6 (the junior college percentage). Thus, 20 + 48 + 6 = 74. Note too that each cumulative percentage equals the cumulative percentage for the preceding value plus the percentage for that value (e.g., 68 + 6 = 74 for junior college or less). The cumulative percentage for the highest value is always 100 percent since all scores must have that value or less.

Cumulative distributions are sometimes more useful than simple frequency or percentage distributions. They answer questions such as what percent of respondents are 40 years or younger, or what percent have incomes of $25,000 or less, or what percent watch television 3 hours or less a day.[3] But cumulative distributions are obviously not applicable to dichotomous variables of any sort—for example, variables like civil disobedience measured as follow one's conscience/obey the law. Moreover, cumulative distributions are usually meaningful only for variables whose values can be ordered. Cumulative frequencies and percentages, therefore, are primarily applicable only to ordinal or interval/ratio variables. For nominal variables like, say, religion or geographic region, it rarely makes sense to use a cumulative distribution since their values have no real ordering to them. We can't speak of "Catholic or less" or "Midwest or less."

After a little experience with data analysis, however, you may find occasional uses for cumulative distributions even for nominal variables. Consider the variable marital status, with categories married, widowed, divorced, separated, or never married coded 1 through 5, respectively. You can use a cumulative distribution to find the number or percentage of cases with scores on the "lowest" four values—i.e., married, widowed, divorced, and separated. These four values constitute the "ever married" cases. The cumulative percentage for separated thus reports the percentage of respondents who have ever been married as opposed to never married.

2.4   Creating Sensible and Well-Formatted Tables

Be sure in your percentage tables—and, indeed, in any analysis—to claim no more precision than your data warrant. Retain only *significant digits*—that is, digits that are reliable and in which you have confidence. In practice this usually means you should round percentages off to either whole numbers or one decimal place. This guideline has exceptions, but not many when working with social scientific data. Percentages with more than one decimal place usually make false claims to precision. Calculators and computers usually give many decimal places (often as many as eight), but most of those digits are not significant.

As a general rule in statistics, keep as many digits as you can while applying formulas and calculating in order to minimize rounding errors during computation. Then round off your final number to

---

[3] The cumulative distributions considered here are all "less than" cumulative distributions. "Greater than" cumulative distributions describe the frequency or percentage of scores that have a given value or *more*.

no more than one more decimal place than you started with. If you begin with whole numbers (as in tallies of cases), round your final numbers (e.g., percentages) to one decimal place or perhaps even whole numbers. Since this is only a general rule, it has exceptions. As always, therefore, *think* about what you are doing and decide how many digits you have confidence in.

Ridiculous extra digits (generally those beyond one decimal place) normally should be rounded to the nearest number. Some examples: Round 21.32 to 21.3 and round 15.66 to 15.7. What about rounding a number ending in 5 like 48.65 or 17.35? A common practice is to round 5s off to the nearest *even* number. Thus, for example, 48.65 is rounded to 48.6, while 17.35 is rounded to 17.4. This "even rule" for rounding is followed in this text. It ensures that in the long run about half of numbers ending in 5 are rounded up and about half are rounded down.

A few words about table format. I have made the percentage tables in the previous section as formal and presentable as I can. Tables of this sort, suitable for term papers, reports, and the like, are described as **presentation-quality**. Although there is no one "official" or "correct" format for percentage tables, the format of tables throughout this text is consistent with the editorial style of the *American Sociological Review* and other American Sociological Association journals. Other disciplines like psychology, political science, and education prefer slightly different formats, and there is variation even within disciplines. Consult a recent issue of a discipline's style manual or major professional journals for format recommendations and examples of properly formatted tables.

Here are some general tips for producing easily read, presentation-quality univariate percentage tables:

- Number your tables with arabic numerals.

- Use a straightforward title that clearly but succinctly identifies the variable described in the table. Also report the data source unless that source is reported in the text accompanying the table.

- Label the left-hand column with the variable name (e.g., Civil Disobedience; Education). Use clear, descriptive names rather than the abridged, sometimes cryptic variable names used in data files.

- Label the right-hand column Percent (or Frequency or Cumulative Percent, as appropriate).

- Be sure that value categories are mutually exclusive and collectively exhaustive (as described in Section 1.6). Every score should be included in one and only one value category.

- Include a Total row that adds up the percentages as a guide to the table's readers.

- Include an (N) row presenting the number of cases on which percentages are based. (Sometimes this row is labeled "Number of Cases.") Presenting N allows a reader to assess the stability of percentages and to calculate the individual frequencies on which percentages are based.

- Be consistent with decimal places. For example, don't round some percentages to whole numbers and others to one decimal place.

- Unless there is good reason to draw attention to frequencies, don't put individual frequencies in the table—just percentages. An interested table reader can recompute any frequency by multiplying N by the percentage and then dividing by 100.

- Keep percentages lined up and right-adjusted. (Most word processors allow you to align decimal points, which is even neater.)

- Don't put % signs after cell entries. They are not needed, they clutter a table, and they are just plain tacky.

- Do not draw vertical lines in a table. They too clutter. To guide a reader's eye and provide definition for the table, draw only a horizontal double line between the title and the column headings and horizontal single lines below the column headings and at the bottom of the table, as shown in tables throughout this text.

- Be very neat. Keep cell entries lined up, align decimal points, keep horizontal lines the same length, and so on.

While style must never be confused with substance, you owe both to readers of your tables. How do you know if your table looks good? A reasonably intelligent person should be able to read the table with minimal effort and no ambiguity. So ask yourself when preparing tables: Could my roommate read this table easily and accurately? (Caution: Apply this test only if your roommate is reasonably intelligent.)

I think the preceding tables look pretty good although I do not claim they are perfect. (Nothing is, of course, except Mozart's music and my wife's chili.) I need to think about this percentage distribution and play with it some more. For example, perhaps Table 2.7 reporting educational percentages has too many categories. Maybe there are too few junior college graduates and respondents with graduate degrees to merit their own separate categories, so maybe they should be combined with the B.A. category. That is, maybe these three categories should be collapsed into a single category.

2.5   Collapsing Variables

Speaking of collapsing categories: Some variables have so many values that it is very helpful to collapse value categories into a more manageable number for frequency or percentage tables. The variable for age of respondent in the General Social Survey, for example, has over 70 values ranging from 18 to 89. Many values have few cases (e.g., there are only five 85-year-olds), and almost six dozen categories are hard to make sense of when summarizing the distribution of this age variable. With so many values, the percentage distribution for age takes over a page to print. A major purpose of descriptive statistics is to condense information to make it more comprehensible. We will find techniques later on that make very good use of all the possible values of variables like age, but too many values get in the way of effective percentage tables. A percentage table over a page long hardly serves to make information more comprehensible. As long as we can do without extensive detail, we can make much more sense of an age distribution collapsed into, say, five or six categories. Depending on our research goals, maybe categories like these would work for the GSS:

          18 – 29
          30 – 39
          40 – 49
          50 – 64
          65 or older

Please notice, not so incidentally, that these categories are mutually exclusive and collectively exhaustive.

Interval/ratio variables often have many values and thus need to be collapsed for techniques like percentaging, but so too do many nominal and ordinal variables. For example, a variable identifying the states in which respondents live (a nominal variable) is usually more useful after being collapsed into perhaps eight or nine major geographic regions such as New England, Mid-Atlantic, and so on. This variable may be even more useful for some purposes if its values are further collapsed into, say, North, South, and West. There are situations, of course, in which we should retain the detail provided by numerous values, but it is often helpful to combine or collapse values even at the loss of some information.

We may also want to collapse categories that have very few cases to form categories with larger frequencies. For example, asked whether they find life exciting, routine, or dull, most General Social

Survey respondents report that life is either exciting or routine. Relatively few—less than 5 percent—report that life is dull. Unless there is good reason to retain the value dull, we may want to collapse dull with routine into a single routine/dull category.

Researchers often collapse categories that offer more detail than is necessary for a given analysis. For example, to measure political attitude, the General Social Survey asks respondents to locate themselves on a seven-point scale with three degrees of liberal and three degrees of conservative, and with moderate in between. That scale works well for some statistical purposes. But for percentage tables it may be useful to combine the three liberal categories and the three conservative categories so that we have three values: liberal, moderate, and conservative.

There are, therefore, several good reasons to collapse values: to create a more manageable number of categories; to handle values with few cases; and to eliminate unnecessary or extraneous detail. But how do you decide how to collapse a variable? How many categories to use? And what categories? These are not easy questions, and reasonable people may well disagree about the best way to collapse a given variable. But here are some guidelines to keep in mind, roughly in order of precedence, from most important to least important.

1. Collapse a variable in a way that makes sense for your research question or objectives. What do you want to find out from the data? For example, it might make sense to collapse Protestants and Catholics into a single category if your research question concerns the differences between Christians and non-Christians. But collapsing Protestants and Catholics makes no sense if you want to compare these two religious groups.

2. Establish categories that are collectively exhaustive and mutually exclusive. That is, make sure that every score is in one and only one collapsed category. There is one exception (and it is a big one) to the collectively exhaustive guideline: You may want to exclude missing data such as Inapplicable, Don't Know, No Opinion, Refused, and No Answer responses. More on this in the next section.

3. Don't collapse a variable so much that you obscure important patterns in the distribution, but also don't have so many categories that the distribution is hard to comprehend. Unless you want to consider details in a distribution, generally try to have no more than six or seven categories—even fewer if it doesn't obscure essential information.

4. Keep variable categories homogeneous and use culturally defined categories as much as possible. In the age example men-

tioned earlier, it makes some sense to have a category for respondents 65 and older since that is more or less the retirement age group in the United States. Likewise, for political party preference in the United States, with categories of strongly Republican, Republican, moderately Republican, and their counterparts for Democrats, it may make sense to collapse the three Republican values into "Republican" and to collapse the three Democrat values into "Democrat."

5. Within all the above criteria and within reason, keep categories of interval/ratio variables the same width. Incomes, for example, may be broken into broad categories each $20,000 wide. Don't make a fetish of equal categories, but keep them equal if you can also adhere to other guidelines for collapsing variables.

6. Within all the above guidelines, follow cultural conventions by creating category widths divisible by powers of 10 for interval/ratio variables. Sensible income categories, for example, might be $0 – 19,999, $20,000 – 39,999, $40,000 – 59,999 . . . . It would be confusing (and dumb) to use bizarre categories like $0 – 14,332, $14,333 – 28,665, $28,666 – 42,997 . . . . We just don't think in those kinds of numbers. Of course, for variables like income that have some extreme values, you will need an open-ended category at one end (or even both ends) of the distribution to capture cases with extreme scores. Income, for example, might have a category such as $100,000 or more to capture high incomes that could go into the millions.

7. Within all the above criteria, establish categories with fairly equal numbers of cases in them if you plan to use the variable in bi- or multivariate cross-tabulations of the sort taken up in Parts II and III of this text. (This guideline helps ensure enough cases for percentaging, a concern we will take up in Chapter 5.)

So what should you do when you collapse a variable? *Think*, that's what! In fact, *think hard* about what categories make sense, what categories best lend themselves to what you want to find out from the data, what categories preserve the most information while making data most manageable. As you will learn in the workbook exercise for this chapter, the MicroCase part of collapsing variables is easy. *Thinking* about what you want MicroCase to do is the hard—but indispensable—part.

For example, consider the General Social Survey variable that reports respondents' years of education. This variable has 21 values ranging from 0 (i.e., no schooling) to 20 (i.e., 8 years of college). To collapse this variable, we could establish the categories 0 – 6 years, 7 – 13 years, and 14 – 20 years of schooling. That would certainly keep categories of equal width. The problem is that such a threefold classification does not capture very well either the variable's distribution or the cultural definitions and social meanings of schooling in America. It does not recognize the very large proportion of Americans who have high school diplomas but have not attended college. And it lumps together 7th-grade dropouts with persons who have a year of college. College dropouts are included with respondents with 4 years of graduate training. The categories are a cultural hodgepodge that would not serve us well.

It's far better, I think, to collapse years of education of Americans into 0 – 11 years, 12 years, 13 – 15 years, 16 years, and 17 – 20 years. Here we have culturally meaningful categories of less than high school, high school graduates, respondents with roughly junior college educations, college graduates, and respondents with graduate educations. There surely are other good ways of collapsing years of education, depending on the research question being addressed or the design of the analysis. It may even be good in some research situations to collapse education still further into just three categories. The important point, once more, is to *think* about what you want to find out and how you can best go about your data analysis.

2.6   Excluding Missing Data

We saw in Section 1.8 that some questions in surveys may not be applicable to certain respondents. There is no sense, for example, asking unmarried respondents how happily married they are. Furthermore, some surveys like the General Social Survey rotate questions so that not every respondent is asked every question. This allows surveys to ask more questions, although of somewhat fewer respondents. A code of –9999 is assigned as a score for cases for which a variable is inapplicable or information on the variable was not collected. And too, some respondents in surveys may not be able or willing to answer certain questions, or they may have no opinion about the question posed. These responses, such as they are, are recorded as Don't Know, No Answer, Refused, No Opinion, and Can't Choose. Likewise, information about cases may simply be unavailable for some cases in aggregate or other nonsurvey data sets. Perhaps the information was never collected. For example, information on defense expenditures may be unavailable for Russia or Iran. Such data is Not Ascertained or Not Available.

Most researchers exclude values like No Answer, No Opinion, and Not Available from an analysis since they don't provide any useful information about cases' scores. There is little reason to include them in the computation of percentages, for example. These kinds of values that we should exclude from an analysis are called *missing data*. Here are values that we usually (although not always) treat as missing data: Don't Know, No Opinion, Can't Answer, No Answer, Refused, Not Applicable, and Not Ascertained.

Sometimes, although not always, the value Other is also regarded as missing data since it is often a residual category with a motley collection of cases that have little in common. The Other category for a variable like religious preference, for example, includes such diverse faiths as Islam, Hinduism, and Buddhism. Whether such a category should be included or excluded in an analysis depends on the research question being addressed. The same is true of categories such as "It Depends" or "Can't Choose." Depending on the research question or goal, it may make sense to include such categories as representing meaningful intermediate positions, or it may make sense to exclude them as not providing useful information. For instance, on the GSS question asking respondents if they think abortion should be legal if the woman wants one for any reason, we might regard Don't Know as indicating an uncertainty that places a respondent between Yes and No. If so, then we would not exclude Don't Know (although we would continue to exclude the value No Answer). No Opinion is similarly ambiguous and may sometimes be included in an analysis as a meaningful value indicating an intermediate position on an issue. Decisions whether to include or exclude values such as Don't Know and No Opinion depend on our research issue and on our semantic interpretation of responses to questionnaire items. There is no cut-and-dried right way to always handle such common research situations. As always, *think* when identifying missing data.

Excluding missing data makes a difference since it changes the total number of cases—N—on which percentages and other statistics are based. Table 2.9, for example, presents the distribution of a General Social Survey variable for opinion about a future world war both with and without missing data categories. Although the relative magnitudes of the percentages remain the same, the percentages themselves increase with missing data excluded. The percent foreseeing a future world war increases from 38.6 to 41.1, and the percent disagreeing increases from 55.4 to 58.9.

Ideally there would be no missing data. We prefer that every case have information on every variable in an analysis. But rarely is that the situation we face. Like missing socks in our laundry, missing data is further evidence that we live in a world that is less than perfect.

Table 2.9. U.S in World War in 10 years,
with Missing Data Included
and Excluded (in percentages)

U.S. in World War?	Missing Data Included	Missing Data Excluded
Yes	38.6	41.1
No	55.4	58.9
Don't Know	4.5	—
No Answer	1.6	—
Total	100.0	100.0
(N)	(1960)	(1841)

2.7   Selecting Subsets of Cases

We often find it useful to restrict an analysis to a particular set of cases. If we are interested in educational achievement in the United States, for example, we may want to restrict our analysis to respondents over the age of 25 since substantial proportions of Americans in their late teens or early twenties are still in school. Or if we study class distinctions among African Americans, we may reasonably choose to limit our analysis to African Americans. If we are interested in variation among the 50 American states in revenues from legal gambling, we may want to restrict our analysis to states that have legalized gambling. A *subset* is such a set of cases selected for analysis based on their scores on some particular variable. Examples of subsets are respondents over age 25, African-American respondents, and states with legalized gambling.

2.8   Pie Charts and Bar Graphs

Much research supports the commonplace observation that human beings grasp information more quickly and retain it longer if the information is presented visually in a chart or graph. This preference for visual information is probably both biological and cultural. Even persons turned off by "statistics" usually like graphic presentations. Consider the success of *USA Today* with its splashy graphics.

There are many kinds of graphing techniques, but the two most common ones for univariate analysis are *pie charts* and *bar graphs*. Figure 2.1 shows a pie chart of the distribution of the variable Civil Disobedience among our 50 respondents. Reading a graph like this is as easy as pie. (Sorry, I couldn't resist.) The whole circle or pie represents the total of all respondents. Slices can represent either percentages or frequencies. In either case, the size of each pie slice is proportional to the percentage or frequency of cases with a given value. The greater the percentage or number of cases with a certain value, the larger its slice of the pie. Since 58 percent of respondents believe people should follow their consciences, their slice is 58 percent of the pie. The 42 percent of cases who believe people should always obey the law have 42 percent of the pie.

Figure 2.1. Attitudes Toward Civil Disobedience

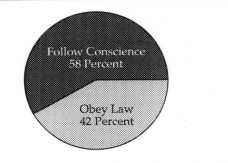

Pie charts are typically used only for variables with less than, say, 8 or 9 values, although there are no hard and fast rules on this. With more values, some pie slices will be too thin to depict very clearly. It's better to use bar charts, described below, for variables with 9 or 10 or more values.

Moreover, pie charts are used primarily for nominal variables or (as above) dichotomous ordinal variables. Multivalued ordinal and interval/ratio variables are usually better displayed using bar graphs that depict the rank-ordering of their values far more clearly. Here we have our first example of level of measurement influencing our choice of statistical techniques. If a variable is nominal or dichotomous ordinal, consider using a pie chart. If a variable is nondichotomous ordinal or interval/ratio or has many values, you will probably want to use a bar graph instead. Figure 2.2 shows one for the frequency distribution of educational levels in our 50-case example.

Like pie charts, bar graphs can display either percentages or frequencies. To state the obvious: The more cases with a given value, the higher the bar. The height of each bar is proportional to the percentage or number of cases with that value. Here in Figure 2.2, for example, there are three times as many college graduates as there are cases who finished junior college, so the B.A. bar is three times as high as the junior college bar. The frequency or percentage of cases with each value can be read off the vertical axis on the left-hand side.

By the way, this kind of bar graph is called a *histogram* if the variable is a continuous ordinal or interval/ratio variable. In that case, the bars should touch to indicate that the variable is continuous. In Figure 2.2, I have treated the units of education completed as discrete, so the bars do not touch.

Figure 2.2. Education Completed

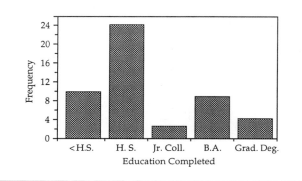

There are several conventions you should follow in creating your own bar graphs:[4]

- Number each graph as a "Figure" followed by an arabic numeral.
- Give the graph a clear but concise title identifying the variable depicted.
- The vertical axis should be approximately 60 to 75 percent the length of the horizontal axis. This ratio gives a uniform appearance to graphs and thus facilitates comparisons. You may need to stretch (quite literally) this guideline for variables like age in years that have many values.
- Keep the scales of graphs constant if you are creating two or more graphs with similar variables (e.g., respondent's education and spouse's education). Using the same scale makes comparisons easier.
- For ordinal or interval/ratio variables, list values from lowest to highest as you go from left to right on the horizontal axis.

[4] If you want to produce presentation-quality graphs, consider using a program specifically designed for graphing. We will soon learn how to produce pretty good-looking graphs using MicroCase, but specialized graphing programs have many more editing features than MicroCase and thus can be used to produce very high-quality graphs.

- Bars should be of equal width so that areas of bars are proportional to their frequencies or percentages.

- There should be spaces between the bars of variables whose values are discrete categories (these educational units, for example). But leave no space between the bars for a continuous variable (like education if it were measured in actual years completed).

- Label the vertical axis with "Frequency" or "Percent" so that a reader knows which it displays.

- The vertical axis usually should begin at zero to avoid distorting perceptions of the areas of the bars. Otherwise, taller bars will seem unduly tall relative to shorter bars and differences between the heights of bars will appear magnified. (You will often see this guideline violated in the mass media, and there are times when you will need to make exceptions yourself. Watch out for violations in the media, however, and be very careful with your own exceptions to this rule.)

These guidelines hit only some of the high points of well-designed graphs. Fortunately, there are many excellent books describing how to create effective graphs. I have listed some of the best in the bibliography at the end of this text.

2.9   Outliers

Bar graphs are particularly helpful for spotting *outliers*—scores on interval/ratio variables that stand alone as unusually high or unusually low. Outliers are not simply low or high scores, but rather are scores that are isolated at the end of a distribution, remote from most other scores. We will learn several statistical procedures that are confounded by outliers, so we need to take them into consideration in analyzing data.

Figure 2.3 displays a bar graph (more specifically, a histogram) for the U.S. General Social Survey variable for hours per day watching television. See any outliers? Sure—those tiny bars on the far right, with scores seemingly unattached to the rest of the distribution. These are respondents who report watching television 18, 20, 22, even an incredible 24 hours a day. Maybe these are folks glued to the Weather Channel or Home Shopping Network even as you read this. More likely, these are respondents who exaggerated their self-reported television watching. We can recognize outliers in univariate frequency or percentage tables too, but they are usually easier to spot in a bar graph. They stand out on the far left or far right.

Always be on the outlook for outliers since they can affect an analysis, often adversely. We'll see examples of these effects in later chapters. When you find outliers, first try to figure out why they occur. Are they mistakes in measurement or preparation of data? If so, we need to correct the errors. Or do outliers suggest some unusual process occurring? That is, are they anomalies—exceptions to the general pattern we expect to find? If so, we need to try to make sense of them, perhaps by revising our theoretical ideas. Or are outliers just flukes, unusual occurrences (like Joe DiMaggio's record 56-game hitting streak) that have no special theoretical implications even though they may disturb our analysis? For TV watching, I suspect that we just have a few unusual cases of very devoted couch potatoes or respondents who exaggerated their TV watching. If so, we may want to delete these cases from our analysis.

Figure 2.3. Hours Per Day Watching Television

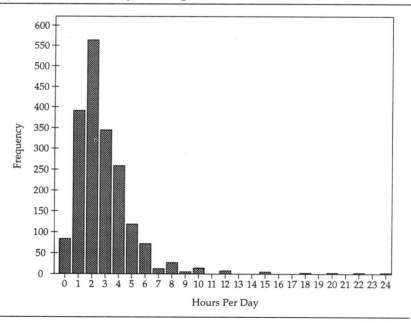

We often exclude outliers in data analyses. We will learn some procedures for doing so in the workbook, and some reasons for doing so scattered throughout this text. But we do not exclude outliers automatically. We need first to determine their meaning and implications for our analysis, and then decide whether their exclusion helps or hinders our reaching our research objectives. We may want to carry out two analyses, one with and one without outliers, and then decide which analysis best answers our research question. My general point:

Thinking is the most important process in statistics. We will learn to handle outliers in due course.

2.10   Mapping Ecological Variables

Ecological data are often continuous interval/ratio variables such as averages, rates, and percentages that take on so many values that they do not readily lend themselves to pie charts, bar graphs, or frequency or percentage distributions. Consider, for example, the fertility rates of the 50 most populous countries. Defined as average number of children that would be born per woman, rates vary from 1.37 in Italy to 7.15 in Uganda, with few countries having the same rate. Therefore, a graph or tabular distribution would not tell us much. A bar graph would show most bars the same height, each indicating a single country with that fertility rate. A table would show a listing of the 50 fertility rates with just one or perhaps a couple cases with any particular value. Of course, we could collapse the variable, and sometimes it is useful to do so. Collapsing, however, always involves the loss of information.

There are, however, other kinds of distributions besides frequency and percentage distributions. Spatial distributions over geographic areas are best displayed on maps. Indeed, maps were among the principal techniques used by early social scientists like Adolphe Quetelet of Belgium and André Guerry of France in the 1820s and 30s. Variables such as fertility rate can be shown on an outline or *area map* with countries colored or shaded to indicate each country's score on the variable. We can thus see how a variable like fertility is distributed geographically around the world.

Figure 2.4 is an area map of fertility rates for the 50 most populous countries. The map displays the high fertility rates in Africa and parts of the Middle East. We find low fertility in Europe and parts of East Asia. These patterns are seen easily only with a map. We would have a hard time noticing them from a simple listing of countries and fertility rates.

Figure 2.4. Fertility Rates (50 Most Populous Nations)

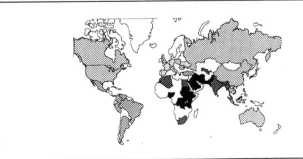

(By the way, white areas on the map are either countries not among the 50 most populous or countries for which information is unavailable.)

Compare the map of fertility rates with the map in Figure 2.5 showing the percentage of population living in urban areas. Hmmm . . . interesting. We see a pattern quite similar to the map for fertility rate. Of course the coloring is reversed, with darker countries on the fertility map usually having lighter colors in the urbanism map. But the patterns are obviously similar. Europe and parts of East Asia are the most urban and have the lowest fertility. Africa and parts of Asia are least urban, just as they also have the highest fertility. The maps certainly do not show identical patterns, but they are fairly similar.

Figure 2.5. Percent Urban (50 Most Populous Nations)

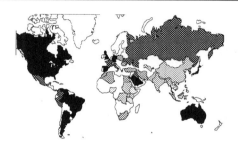

Area maps like Figures 2.4 and 2.5 are sometimes misleading, however, because large units are far more prominent than small units. Our eyes—and brains—naturally pay more attention to vast countries like the United States and Russia than to small countries like Japan and European nations. Sometimes we get a better visual sense of the spatial distribution of a variable by using a *spot map* like the one in Figure 2.6, showing percent urban. This map presents the same information as the map in Figure 2.5, but depicts each country's urbanism by a spot rather than coloring the entire country.

Figure 2.6. Spot Map of Percent Urban (50 Most Populous Nations)

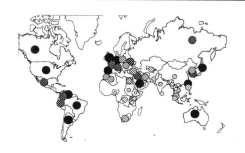

The size of each spot indicates each country's urbanism rate. The bigger the spot, the higher the percentage urban. Note that spot sizes are independent of the geographic area of states. In fact, little Japan and small European countries, highly urban, now stand out with large dots.

In Chapter 10 we will learn to analyze the connections between these variables more efficiently, but certainly maps can give us some sense of the similarities in patterns. Mapping is an important tool for describing an ecological variable and comparing its geographic distribution to other ecological variables. Workbook Chapter 2 will help you create your own maps quite easily and will introduce you to some additional features of MicroCase's mapping procedure.

2.11   Summing Up Chapter 2

Here is what we have learned in this chapter:

- Frequency and percentage distributions summarize the pattern of scores for variables.

- Percentages standardize frequencies to base 100, thus making data patterns easier to interpret and compare.

- Percentages should be used with caution when the total number of scores is small (say, less than 30).

- Cumulative distributions show the number or percentage of scores with a given value or less.

- Meaningful cumulative distributions usually can be found only for ordinal or interval/ratio variables.

- Tables presented to others should be readable and follow a conventional format.

- When collapsing values of a variable, we should think about the most useful way to collapse values, given the research question we are addressing and the way we have designed our analysis.

- We usually should regard values such as Inapplicable, Don't Know, No Opinion, Can't Choose, No Answer, Refused, Not Applicable, and Not Ascertained as missing data to be excluded from an analysis. Sometimes we also treat Other as missing data.

- When we make semantic sense of values such as Don't Know and No Opinion, we should include cases with such values in an analysis.

- Sometimes we need to select a subset of cases on the basis of their scores on one or more specified variables.

- Frequency and percentage distributions can be displayed visually in pie charts and bar graphs.
- Bar graphs are usually preferable to pie charts for ordinal and interval/ratio variables and for nominal variables that have many values.
- We need to watch for outliers—extreme values that may distort a statistical analysis.
- Area and spot maps display spatial patterns of ecological variables.

Key Concepts and Procedures

Ideas and Terms

frequency distribution	collapsing variables
raw data	missing data
frequency	subset
frequency table	pie chart
percentage	bar graph
proportion	histogram
percentage table	outlier
cumulative percentage	mapping
significant digits	area map
presentation-quality table	spot map

Symbols

f
N
F

Formulas

$$\text{Percent} = \frac{f}{N}(100)$$

$$\text{Cumulative Percent} = \frac{F}{N}(100)$$

ANALYSIS WRITE-UP 1: PERCENTAGES, DISTRIBUTIONS, GRAPHS, AND MAPS

◆   ◆   ◆   ◆   ◆

You may find models of statistical write-ups and accompanying tables helpful as you present results of your own analyses. I will, therefore, offer a few examples of write-ups after most chapters. These models are certainly not the only way to describe analyses. Stylistic and format conventions are fairly loose throughout most of the social sciences, and your instructor may have his/her own preferences on write-ups for you to follow. You may also find a good published guide to presenting statistical results helpful. I have listed several in the bibliography at the end of this text.

You will find this text's examples of write-ups straightforward—no frills/nothing fancy. Simplicity is a virtue in presenting research results. Bear in mind that each write-up offered in this text is necessarily out of context. Embedded in an actual paper or report, each write-up would almost certainly be preceded by a description of the data set, including the source of data, the sample or population, and the measurement of variables. Those methodological discussions would usually be preceded by a statement of the research problem and/or theoretical context, any specific hypotheses being tested, and a review of the relevant literature. And, too, each finding would be followed by discussion of the analysis' theoretical or applied implications and probably a short conclusion at the end. I have omitted all these important parts of a paper or report to focus on the matter at hand: statistical findings.

In several write-ups I offer variations on table formats to better accommodate multiple analyses in a single table. You will see throughout my examples of write-ups that it is customary to indicate where a given table goes by inserting "Table X About Here" after the first paragraph that refers to the table. Tables themselves go at the end of the paper or report.

Some of the variables in Analysis Write-ups are in the data files accompanying this text and workbook, especially the General Social Survey data. However, I have not restricted myself to these variables and freely draw on variables from the complete General Social Survey. All Write-ups, of course, exemplify statistical procedures carried out by MicroCase.

By the way, my comments about the write-ups are in italics like this and are not part of the write-up.

A. *Here is a write-up describing the percentage distribution of a single variable.*

The 1996 General Social Survey asked respondents to describe their social class as lower, working, middle, or upper class. Table 1 presents the distribution of subjective class identification. Overwhelmingly, Americans regard themselves as either working or middle class. About 45 percent of respondents report that they are middle class, and the same percent identify themselves as working class. Less than 6 percent of respondents regard themselves as lower class, and about 4 percent as upper class.

TABLE 1 ABOUT HERE

Include the table at the end of the paper or report:

Table 1. Subjective Social Class
(in percentages)

Class Identification	Percent
Upper Class	4.2
Middle Class	44.9
Working Class	45.3
Lower Class	5.7
Total	100.0
(N)	(1829)

B. *A single table often works well for several univariate analyses of related variables. The write-up highlights selective aspects of particular univariate distributions without necessarily discussing all. Note that Table 2 below displays distributions horizontally. It works better that way when multiple variables are displayed in a single table. As you become more comfortable with statistics, you will want to vary your statistical tables to present information as clearly and efficiently as you can.*

General Social Survey respondents were asked whether they have a great deal of confidence, only some, or hardly any confidence in each of several social institutions, both public and private. Table 2 describes confidence in these institutions. Respondents report considerably less confidence in government or media than in financial institutions, religion, or education. Indeed, little more than one in ten respondents express great confidence in government or media, whereas about a quarter report great confidence in financial, religious, and educational institutions. Likewise, substantially greater proportions of respondents register hardly any confidence in government, television, or the press.

TABLE 2 ABOUT HERE

Include the table at the end of the paper or report:

Table 2. Confidence in Selected Social Institutions
          (in percentages)

Institutions	Great Deal	Only Some	Hardly Any	Total	(N)
Financial Institutions	25.4	57.7	25.4	100.0	(1873)
Religion	26.3	53.7	19.1	100.0	(1831)
Education	23.2	58.4	18.4	100.0	(1900)
Federal Government	10.5	46.4	43.1	100.0	(1856)
Press	11.0	49.2	39.9	100.0	(1867)
Television	10.4	47.2	42.5	100.0	(1881)

C. *Graphs are often a good way to present univariate distributions. Be sure to consider your options. Table or graph? Pie chart or bar graph? Map? Choose whatever options allow you to present the information you want to get across to your reader or audience. Here I have chosen a bar graph.*

The two-child family is normative in the United States. A clear majority of 1882 General Social Survey respondents report two as the ideal number of children for a family to have. Much less than half as many regard three children as the ideal, with decreasing proportions treating even more children as desirable. Very few Americans endorse no children or only one child. Figure 1 displays these results.

FIGURE 1 ABOUT HERE

Include the figure at the end of the paper or report:

Incidentally, I prepared this figure with a specialized graphics program. Note that values 0 through 7+ form a continuous variable and so are contiguous. The bar for Number Wanted represents a discrete value and so is separated out.

Figure 1. Ideal Number of Children

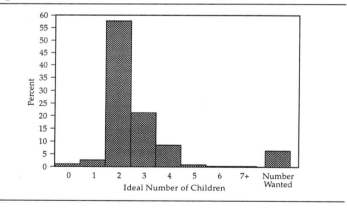

D. Maps work well to describe the spatial distribution of variables.

Social scientists define a sex ratio as the number of males per 100 females in a population. Figure 2 displays variation in sex ratios across the United States. A diagonal swath of states with an underrepresentation of men runs from New England and New York down to Mississippi and Louisiana, with much of the upper Midwest also having low sex ratios. On the other hand, the Rocky Mountain states, the far West, Alaska, and Hawaii have a relative dearth of women.

FIGURE 2 ABOUT HERE

Include the figure at the end of the paper or report:

Figure 2. Sex Ratios of 50 States

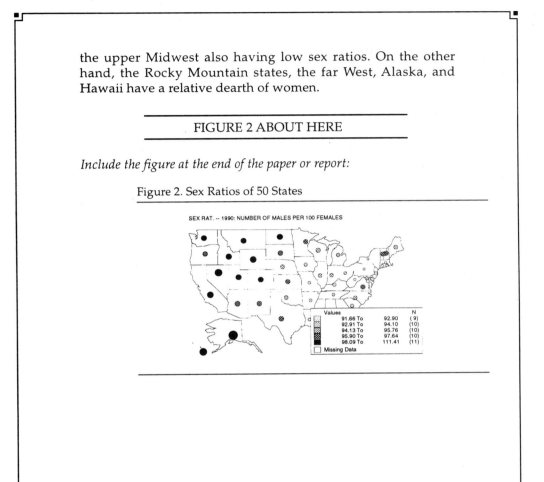

SEX RAT. -- 1990: NUMBER OF MALES PER 100 FEMALES

Values		N
91.66 To	92.90	(9)
92.91 To	94.10	(10)
94.13 To	95.76	(10)
95.90 To	97.64	(10)
98.09 To	111.41	(11)
Missing Data		

CHAPTER 3
Averages

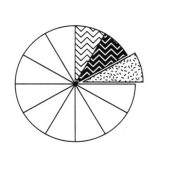

Percentage distributions and visual displays such as graphs and maps are certainly helpful in summarizing information about a variable. These statistical techniques go a long way toward making information more manageable and easier to communicate. But we can summarize univariate information even more concisely (although with still greater loss of detail) by computing *averages* or, in statistical jargon, ***measures of central tendency***. (Statisticians prefer the technical term "measure of central tendency," but we will use the more familiar "average" here.) An average is a typical or representative value for a set of scores. Like percentaging, averaging is something you learned in elementary school or junior high, but we will review averages here in case you are a little rusty on them.

After this chapter you will be able to:

1. Calculate and interpret modes, medians, and means.

2. Explain important properties of the mode, median, and mean.

3. Explain and recognize conditions under which each type of average is appropriate.

4. Explain the effects of extreme scores on means.

5. Explain the lack of effects of extreme scores on modes and medians.

6. Explain what a sum of squares is and calculate sums of squares.

7. Recognize unimodal and bimodal distributions.

8. Identify negatively and positively skewed distributions and their effects on medians and means.

3.1   Mode

There are actually many kinds of averages, some of which are quite esoteric. (Ever hear of geometric or harmonic means?) However, three kinds of averages are particularly useful—the mode, median, and mean. Let's consider each in turn.

A *mode* (sometimes abbreviated *Mo*) is the most frequently occurring score on a variable. In the United States (and most other countries), the modal gender is female because there are more women than men. The modal years of education completed by Americans is 12 because there are more people with exactly 12 years education than any other value. A glance back at the bar graph of daily hours of TV watching in Figure 2.3 of Section 2.9 will show you that the mode of that variable is 2—more respondents watch television 2 hours a day than any other number of hours.

When clearly there is only one score that is most common or the general distribution shows only one hump, we say that the variable is *unimodal*. A bar graph is the best way to identify one dominant score or a pronounced hump in a distribution. General Social Survey respondents' hours watching TV per day (see Figure 2.3) is unimodal—there are many more scores of 2 than any other score. GSS respondents' notions of the ideal number of children is also strongly unimodal. Over half (55 percent) of respondents report that two children is ideal, and its bar towers over others. Reflecting demographic trends, the variable age does not have a single dominant score, but its distribution shows a gradual increase to the late 30s, and then a gradual decrease into old age. Age too has a unimodal distribution.

Sometimes, however, a bar graph (or histogram) of an ordinal or interval/ratio variable shows a camel-shaped distribution with two humps. Then we describe the variable as *bimodal*. The two humps do not have to be exactly the same height. They just each need to be roughly the same height and quite a bit higher than the bars for other values for us to say a variable is bimodal. Figure 3.1 displays the bar graph for a bimodal General Social Survey variable reporting respondents' number of children. Obvious modes appear at the 0 and 2, each of which has over 25 percent of all cases. These two bars tower over the others.

Sometimes, of course, a variable's distribution is quite flat with no particular score having a large proportion of cases. The General Social Survey variable for the fundamentalism/liberalism of respondents' religions is this kind of flatly distributed variable. Scores are distributed about equally across the fundamentalist-moderate-liberal categories, with no single category having a particularly large number of cases. In this situation, we note that the variable has no mode. If we recognize a mode but it is not very pronounced, we describe the distribution as weakly modal.

Figure 3.1. Number of Children

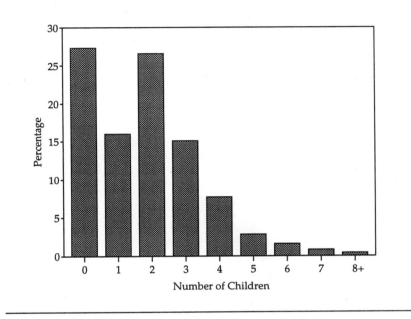

There is no formula for the mode. We find the mode simply by identifying the value that occurs most often among scores. We can do this easily with a bar graph or with a frequency or percentage table.

The mode depends only on differences among frequencies of scores. It does not involve either the rank-ordering of values or units of measurement. Thus, we can find modes for variables at any level of measurement—nominal, ordinal, or interval/ratio. The mode is the only kind of average that we can usually use for nominal variables.

The mode, however, is often not very useful for continuous interval/ratio variables that distribute cases fairly evenly over many values because no single value then stands out as having a much larger frequency than others. For example, the continuous interval/ratio variables (e.g., murders per 100,000 population) found in data pertaining to the 50 American states have few cases with the same score, and thus do not have modes that are of any interest to us. For such variables, the mode provides a very poor indicator of a typical or average score.

3.2   Median

The *median* (sometimes abbreviated *Md*) is the value that divides an ordered set of scores in half. "An ordered set of scores" refers to scores arranged in rank-order from the lowest to the highest. The median is the point at which half the scores are lower and half the scores are higher. In other words, it is the midpoint or center of the ordered scores. Be careful: The median is the midpoint of *scores*—that is, actual measurements—not possible values.

Finding the median is especially easy if there is an odd number of scores:

1.   Put the scores in order from lowest to highest.

2.   Find the middle score.

The value of that middle score is the median. We can find the position of the median score in an ordered distribution with the formula $\dfrac{N + 1}{2}$, where N is the number of cases. For example:

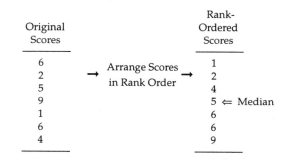

The median is 5. The median 5 is the fourth score $\left(\dfrac{N + 1}{2} = \dfrac{7 + 1}{2} = 4 \right)$ in the ordered set of scores. There are the same number of scores (three of them) less than 5 as there are greater than 5.

If there is an even number of scores, finding the median is a little harder . . . but not much. Here is what to do:

1.   Put the scores in order from lowest to highest.

2.   Find the *two* middle scores.

3.   Average the two middle scores by adding them and then dividing by 2.

The result is the median. Here is an example using this set of six scores (already rank-ordered from low to high):

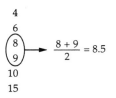

$$\frac{8+9}{2} = 8.5$$

The median is 8.5. Half the scores are lower and half the scores are higher than 8.5.

It doesn't matter if there is more than one of the same score in the middle of a distribution. Since all have the same value, that value is the median. If, for example, there were three more scores of 8 in the distribution above (so that scores were 4, 6, 8, 8, 8, 8, 9, 10, 15), the median would be 8.

Nor does it matter if an ordinal variable has alphabetic rather than numeric values. Its scores can still be rank-ordered and we can find the median. Here, for example, is how to find the median for the variable class identification, with values lower, working, middle, and upper:

Original Scores		Rank-Ordered Scores
Middle	→ Arrange Scores in Rank Order →	Lower
Working		Working
Lower		Working
Upper		Middle ⇐ Median
Middle		Middle
Middle		Middle
Working		Upper

Yes, finding the median is very easy. But now I will add a small complication created by the kind of data we work with in actual research. Social scientists often use variables that are conceptually continuous in that theoretically they can have any value, but that have been measured using discrete, integer (i.e., whole number) scores. Many attitudinal variables in social surveys are like this. For example, the General Social Survey asked respondents how strongly they agreed or disagreed with this statement: "Immigrants take jobs away from people who were born in America." The five response choices were coded:

1   Agree strongly
2   Agree
3   Neither agree nor disagree
4   Disagree
5   Disagree strongly

Respondents in effect round their scores to the nearest whole number. Even if a respondent really agreed a little more strongly or a little less strongly than "agree," the respondent would still get a code of 2. Conceptually, scores on this variable may range from .5 to 5.5, and a respondent's "true" location may be anywhere along that continuum. Theoretically, a respondent could have a score of, say, 3.2, or 4.8, or any other value between .5 and 5.5. The GSS researchers, however, have sensibly worded this item in a way that, in effect, has respondents round their "true" scores to the nearest whole number. Still, the "true" scores of respondents who answered "agree" and were coded 2, for example, may be anywhere between 1.5 and 2.5, the upper and lower limits of the interval within which scores round to 2.

We usually calculate the median by conveniently assuming that scores in the category containing the median are equally spaced throughout that category's interval. Within that interval, we then find the score that divides the total distribution in half. This procedure gives us a more refined, sensitive median than does just crudely using the integer value of the category containing the median.

For example, here are the frequency and cumulative frequency distributions for the variable measuring respondents' attitude toward immigrants and jobs:

Value	f	F	
1	170	170	
2	446	616	
3	299	915	⇐ Median is the value of the 641st score
4	301	1216	(i.e., somewhere among the 299 scores
5	65	1281	of 3)
(N)	(1281)		

Recall from Section 2.3 that F is the symbol for cumulative frequencies. The cumulative frequency distribution helps us find where the median is located. Since there are 1281 scores, the median is the value of the 641st score. (There are 640 scores less than and 640 scores greater than the median.) We can tell from the cumulative distribution that the 641st score lies somewhere among the 299 scores of 3.

To compute the median, we assume the scores for these 299 respondents in the interval containing the median are spaced equally between 2.5 and 3.5, the lower and upper limits for value 3. We interpolate to locate the median in this interval. Interpolate means to find the point in the 2.5 to 3.5 interval that corresponds to the 641st score.

The location of the median is represented graphically in Figure 3.2. Imagine cases distributed along the entire continuum, and equally spaced within each interval of the continuum. Thus, the 299 cases with scores of 3 are distributed at equal intervals within the range from 2.5 to 3.5. The oval shows a blow-up of the portion of this interval that contains the median. There it is—the 641ˢᵗ score.

Figure 3.2. Interpolation of the Median

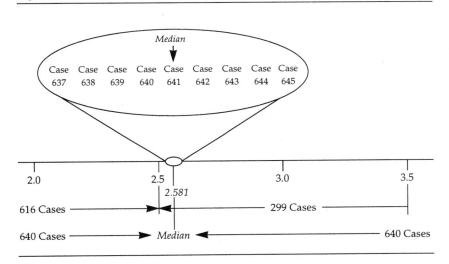

OK . . . but how do we interpolate? By using this formula:

$$Md = L + \left(\frac{\frac{N}{2} - F}{f} \right) (i)$$

where Md = median

L = lower limit of the interval containing the median

N = total number of scores

F = cumulative frequency of scores less than the interval containing the median

f = number of scores in the interval containing the median

i = width of the interval containing the median (i.e., the interval's upper limit minus its lower limit)

For attitude toward immigration and jobs:

$$Md = L + \left(\frac{\frac{N}{2} - F}{f} \right) \text{(i)}$$

$$= 2.5 + \left(\frac{\frac{1281}{2} - 616}{299} \right) \text{(1)}$$

$$= 2.5 + \frac{24.5}{299}$$

$$= 2.5 + .081$$

$$= 2.581$$

$$= 2.6$$

In other words, if the 299 scores of 3 were equally spaced in the interval between 2.5 and 3.5, the median would be 2.581, or 2.6 rounded off. The median lies toward the lower end of the interval for value 3 (neither agree nor disagree). I think you will agree this is a more refined estimate of the median than the crude value of 3 would be.

Since calculation of the median requires that scores be rank-ordered, we can find the median only for ordinal or interval/ratio variables. There is no median for nominal variables because scores for nominal variables cannot be put into rank-order. Any median for nominal variables like gender, race, or religion would be meaningless. So I repeat: There is no median for a nominal variable!

Note that extreme low or high scores do not affect the median because the median describes only the score at the center of an ordered distribution. For example, all the following sets of scores have the same median—55:

Set A	Set B	Set C	
51	1	51	
52	52	52	
54	54	54	
55	55	55	⇐ Median
56	56	56	
56	56	56	
59	59	590	

We will see in the next section that this independence from the effects of extreme scores is an important advantage of the median.

You are usually wise to round the median to one decimal place for variables like those in the General Social Survey. However, there are no hard and fast rules for decimal places, so you will have to think carefully about the appropriate number of decimal places for any median you find.

Be sure to exclude missing data—Don't Know, No Answer, and such—from your calculations of medians. Such values usually provide little or no information and so are best left out of an analysis. Besides, inclusion of such data would reduce an ordinal or interval/ratio variable to a nominal variable since missing data values are inherently unordered.

You have probably noticed that it doesn't matter if the median represents no actual score or even possible score. The median is still the midpoint of an ordered set of scores. The statistically ignorant are amused when they encounter an "impossible" median. They snicker when told that the median size of American households is 2.7. "How can that be?" they chortle. "Is someone pregnant?" But now you know what a median is and that it need not correspond to any actual score, so no longer will you chuckle at such dumb humor.

3.3   Mean

Finally, let's define the mean. The *mean* is the arithmetical average found by dividing the sum of all scores by the number of scores. Real simple:

1.   Add all the scores.

2.   Divide by the number of scores.

Here is an example using this set of six scores: 4, 8, 10, 11, 9, 6. To find the mean, add all the scores: $4 + 8 + 10 + 11 + 9 + 6 = 48$. Then divide that sum by the number of scores: $\frac{48}{6} = 8$. The mean is 8.

Yes, I know you have been calculating this kind of average for years. Still, a formula for computing the mean for sample data is helpful as a shorthand way of explaining how the mean is calculated:

$$\overline{X} = \frac{\Sigma X_i}{N}$$

where $\overline{X}$ = mean

X_i = score of the ith case

N = number of scores

Some unavoidable notation: $\overline{X}$ is pronounced "X-bar." Statisticians use $\overline{X}$ to refer to the mean of a *sample*. They use μ (pronounced myōo) to represent the mean of a *population*. In general, Roman letters are used for sample statistics and Greek letters for population parameters. (Recall from Section 1.3 that information pertaining to a sample is called a statistic and that information pertaining to a population is a parameter.) So, for population data, the mean's formula becomes

$$\mu = \frac{\Sigma X_i}{N}.$$

The subscript i in X_i is just a way to designate individual scores. It is sort of a generic score. X_1 is the first score, X_2 is the second score, and so on. Thus, X_i is the score of an arbitrary i^{th} case. It does *not* imply that the scores are in any particular order. Σ is the uppercase Greek letter sigma. We will see it often in this text. Statisticians use Σ to mean add all the things that come after it, so ΣX_i means add all the individual scores.

In the simple example above, the mean—8—happens to be an actual value. But like the median, the mean need not equal any actual or even possible value. A class's mean score on a true-false test, for example, might be 77.8 even though no student could actually get such a fractional score.

It is usually good practice to round the mean to one decimal place for most data of the kind we use in the social sciences. But whatever data you use, *think* about the significance of the digits in your calculations. Do not use decimal places that make the mean appear more precise than the data warrant. And by the way, be sure to exclude missing data—Don't Know, No Answer, and so on—when you calculate means. Such data present no information that can be used in calculating means.

Be careful in your interpretation of means for ecological data such as data for countries or U.S. states. The mean for an ecological variable treats each ecological unit as a single case and describes the average for the ecological units, not for the total population of individuals who may comprise the units. For example, the mean for years of education for the 50 states gives equal weight to California and Wyoming despite large differences in their population sizes. The mean for the 50 states, then, reports the average of the 50 states, not the average education of all Americans. The same is true for medians, but misinterpretation of means is a more common pitfall that we have to be careful to avoid.

3.4   Properties of the Mean

Watch for particularly low or particularly high scores when calculating a mean. Certainly watch for outliers. Very low or very high scores may produce a mean that is not very typical of most scores, especially when the number of cases is small. Imagine if the score of 6 in the previous section's example were 600—certainly an outlier. With scores of 4, 8, 10, 11, 9, 600, the mean would be 107, which is not typical or representative of what the six scores "really" are. The median— 10.5—works better as an average.

Or think of the average income on Gilligan's Island. Millionaire Mr. Howell's income grossly exaggerates the average income of Gilligan's Islanders if the mean is used. The median is far more representative of the incomes of Gilligan and his pals on the ill-fated *Minnow*. Mr. Howell's million-dollar income counts as much as 100 Gilligans making 10 grand a year. The same is true on the mainland. It takes 100,000 computer programmers making $50,000 a years to balance Microsoft's Bill Gates' $5 billion a year. That's why the U.S. Bureau of the Census almost always reports median rather than mean incomes.

Another example with real data: Among the 50 American states, the percent of the population of Asian ancestry has an obvious outlier, Hawaii with 61.8 percent Asian. The next highest score is 9.6 percent for California. This variable's mean of 2.8 percent with Hawaii included drops to 1.6 with Hawaii excluded. The latter mean is far more representative of a typical score among the 50 states.

I pointed out in Section 2.9 that when you encounter an outlier, first try to determine why it occurs. Then look at its impact on your analysis—in this case, on the mean. If its impact is substantial (as in the examples in the last three paragraphs), consider either excluding the outlier from your analysis or using the median rather than the mean since the median is not affected by outliers.

Yes, watch for outliers when using means. In fact, watch for distributions that have pronounced tails toward either low scores or high scores even if there are no outliers as such. Since the mean takes all scores into account, these tails may make the mean a poor representative of a "typical" score. Better in such situations to use the median.

But enough about difficulties with the mean. The mean has some very good characteristics too, including a most interesting and important property: If we subtract the mean from each score and add all these differences, the sum is always 0. Stated more succinctly: The sum of deviations of scores from the mean is 0. Algebraically: $\Sigma(X_i - \overline{X}) = 0$ for sample data; $\Sigma(X_i - \mu) = 0$ for population data. Some of the differences are positive and some are negative, and when added they exactly cancel each other.

This property of the mean is of much theoretical and practical interest. The mean "balances" a distribution, even in a literal sense. If cases were weights (say, 1-ounce bars) distributed along a beam at points corresponding to their scores, the mean is the point that perfectly balances the cases. The upper teeter-totter in Figure 3.3 shows this balance graphically for the six scores used above. The three weights on the right are balanced by the two weights farther out on the left. If you try to balance this distribution at the median, it will tip over, as shown in the lower teeter-totter. The lowest scores are farther out and thus tip the teeter-totter down on the left. I will refer to this characteristic of the mean several times throughout this text, so please keep it in mind.

Figure 3.3. The Mean Balances a Set of Scores

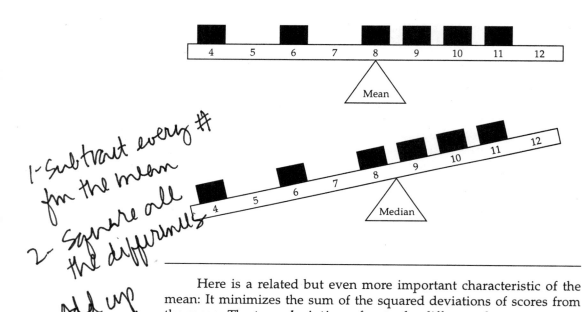

1- Subtract every # from the mean

2- Square all the differences

3- Add up the squared differences

Here is a related but even more important characteristic of the mean: It minimizes the sum of the squared deviations of scores from the mean. The term *deviation* refers to the difference between a score and the mean. So, if we subtract the mean from each score, square each of these differences, and then add the squared differences, we get a sum smaller than we would get for any other number. Algebraically: For sample data, $\Sigma(X_i - \overline{X})^2$ is a minimum; for population data, $\Sigma(X_i - \mu)^2$ is a minimum. No number would produce smaller sums in these expressions than the mean produces.

This expression—$\Sigma(X_i - \overline{X})^2$ for sample data and $\Sigma(X_i - \mu)^2$ for population data—is so important in statistics that it has its own name. It is called a *sum of squares*. This concept is so useful that I want to say again what it is: The sum of squares is the sum of squared deviations from the mean.

As an example, here are the six cases, with $\overline{X} = 8$, used in the previous section:

X	$X_i - \bar{X}$		$(X_i - \bar{X})^2$
4	$4 - 8$ =	-4	16
8	$8 - 8$ =	0	0
10	$10 - 8$ =	2	4
11	$11 - 8$ =	3	9
9	$9 - 8$ =	1	1
6	$6 - 8$ =	-2	4
Sum		0	34

The sum of squares is 34. No other number could be substituted for the mean (8) that would produce a lower sum of squares.

We will make much use of this sum of squares concept later on when we learn analysis of variance and regression and correlation techniques. Please do keep it in mind.

3.5   The Mean for Dichotomous Variables

Note that, strictly speaking, the mean can be calculated only for variables measured at the interval/ratio level. Strictly speaking, it does not make any sense, for example, to calculate mean religion or mean social class. Religion is a nominal variable; it is not measured with any standard unit of measurement. Its values—Protestant, Catholic, Jewish—cannot be added. Similarly, if social class is measured at the ordinal level (e.g., lower, working, middle, and upper classes), a mean, strictly speaking, cannot be calculated. The reason is that (strictly speaking) the values of an ordinal variable cannot be added in any meaningful way because there are not necessarily equal intervals between values. We cannot add the values lower, working, middle, and upper classes.

I used "strictly speaking" several times because it is sometimes useful to *act* as if numbers assigned to values of an ordinal variable have arithmetical meaning, and then to compute a mean using those numbers. In the previous section we found that this was the case for the median, and it is also true for the mean. With social class, for example, it may be useful to assign the number 1 to lower class, 2 to working class, 3 to middle class, and 4 to upper class much as we did in calculating an interpolated median in Section 3.2. We can then calculate a mean using these numbers as scores and refer to the average (i.e., mean) social class. After all, we can sensibly interpret this mean. A social class mean of 2.7, for example, indicates that the average social class of respondents is between working class and middle class, but a little more toward middle class.

This practice of treating ordinal variables as if they are interval/ratio is condemned by statistical purists. But a Spanish proverb says that you can't make a beautiful omelet without breaking eggs, and so too you sometimes can't carry out a beautiful analysis without breaking a few rules . . . *as long as you know what you are doing.* Sometimes we can make better sense out of data if we compute means for ordinal variables. Always, however, we need to be extremely cautious in interpreting these means. As a statistical novice who may not be quite sure yet what you are doing, you should avoid computing means for ordinal variables, but I do want you to know that many social scientists use means with ordinal data. As you become more familiar with statistics and data analysis, you may find it useful to do so yourself (even this semester).

But what about nominal variables? Do some social scientists also assign numbers to the values of a nominal variable like religion and then calculate a mean? The answer is *sometimes* if the variable is dichotomous and an emphatic *NO* if the variable has three or more values. Even we statistical nonpurists don't do that. But if the values of a *dichotomous* variable, even a nominal one, are coded 0 and 1, then the variable's mean equals the proportion of cases coded 1. Suppose, for example, that males are coded 0 and females are coded 1 on the variable gender, and that we have the following five cases:

Gender	Code
Female	1
Male	0
Female	1
Female	1
Male	0
	$\Sigma X_i = 3$

Note that .60 of the cases (three to five) are female. Note too that $\overline{X} = \dfrac{\Sigma X_i}{N} = \dfrac{3}{5} = .60$. Yes, the mean equals the proportion female. This works every time for dichotomous variables as long as scores are coded 0 and 1. In fact, although arithmetical adjustments are necessary, the same principle applies if values of a dichotomous variable are coded any two consecutive numbers. With variables coded 1 and 2, for example, the mean minus 1 equals the proportion of cases coded 2. More generally, the mean minus the lower code equals the proportion designated with the higher code.

This use of dichotomous variables, even nominal ones, is a thread running through the fabric of statistics. In Chapter 10 we will examine the relationships between dichotomous nominal variables using regression and correlation techniques. In Chapter 12 we will transform nondichotomous variables into so-called dummy variables coded 0 and 1 in ways that have useful applications. But more on all this later on. For now, let's not compute means for nominal variables except dichotomous ones, and we'll be especially cautious even with those.

3.6 Which to Use—Mode, Median, or Mean?

We have reviewed three kinds of averages. But which one to use—the mode, median, or mean? The answer depends on a variable's level of measurement, its distribution, and what we want to find out about the variable. Let me offer some guidelines for selecting an appropriate average.

As I pointed out in the previous section, we can (strictly speaking) calculate the mean only for interval/ratio variables. Of course, we can also calculate the mode and median for interval/ratio variables. But other considerations aside (I'll get to those in a bit), we usually prefer the mean for interval/ratio variables. After all, only the mean uses all scores in its calculation and, therefore, the mean takes advantage of more information than does either the mode or the median. We might as well make use of whatever information we have, so other considerations aside for the moment, the mean is usually preferable for interval/ratio variables.

The median is usually the average to choose for ordinal variables since we generally should not compute the mean for such variables (although it is sometimes done, as I noted in the previous section). The median at least uses information about the middle score(s) in a distribution, which is more than the mode does. So all else being equal, use the median for ordinal variables.

That leaves the mode as the preferred average only for nominal variables. After all, we don't have much real choice for nominal variables since their low level of measurement does not permit use of either the median or the mean. Sure, as we have just seen, we can compute a mean for dichotomous nominal variables coded 0 and 1 and interpret that mean as the proportion of cases coded 1. But that doesn't help us with nondichotomous nominal variables. Usually we are stuck with using the mode for nominal variables.

But I don't mean to disparage the lowly mode. Sometimes the mode is useful, especially if we need to know the most likely or commonly occurring score. By definition, more scores occur at the mode than any other value. That's not always true of the median or the mean. The mode, therefore, provides the single "best guess" of a score, and thus is useful if we need a prediction of the most likely actual score. If you had to guess exactly what some unknown score would be, your best guess would be the mode. After all, more cases have that score than any other score. We will make use of this characteristic of the mode when we consider measures of association in Chapter 7. Moreover, the mode is sometimes useful as a supplement to the median or mean for ordinal or interval/ratio variables to give some indication of the shape of a distribution of scores.

Speaking of the shape of a distribution, the mean and the median have the same value in a symmetric distribution—that is, a distribution in which the right half is a mirror image of the left half. And if a symmetric distribution is also unimodal, all three averages—the mode, median, and mean—have the same value. The most frequently occurring value also splits the distribution exactly in half and is the arithmetical average. You can see this visually in the symmetric distribution in Figure 3.4.

Figure 3.4. Symmetric Distribution

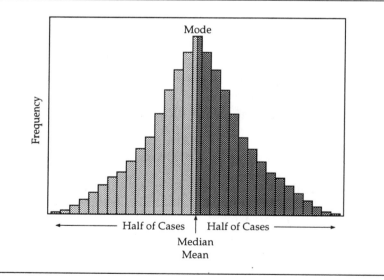

In nonsymmetric distributions, however, the three measures of central tendency take on different values. The median and mean are affected differently by a variable's *skewness*—that is, the extent to which the distribution of a variable has more cases in one direction than the other. A distribution may be skewed to the left by unusually low values. Because the mean is affected by all scores, the low scores "pull" the mean in their direction so that the mean is smaller than the median in a distribution skewed to the left. We describe such a distribution as *negatively skewed*. On the other hand, a distribution skewed to the right—*positively skewed*—has a mean larger than the median since the high scores pull the mean in their direction.

Figure 3.5 shows a graphic representation of the effect of skewness on the mean and the median. You can see that the median always lies between the mode and the mean in a unimodal skewed distribution.

Thus, we need to be careful about distributions that are highly skewed because of either outliers or a "natural," smooth skewing of scores. As I noted before, the mean is not a very typical or representative value in such situations. For highly skewed distributions, the median is usually preferable to the mean even for interval/ratio variables because the median is far more resistant to extreme scores. If outliers are present, consider either excluding them or using the median, or else including the outliers but using the median to represent a typical score.

Figure 3.5. Effects of Left Skewness and Right Skewness on Mean and Median

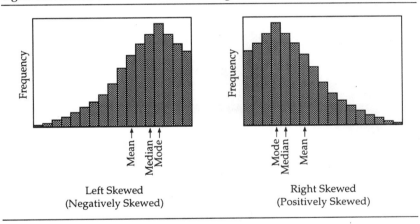

Left Skewed
(Negatively Skewed)

Right Skewed
(Positively Skewed)

Incidentally (but usefully), here is how generations of students have remembered to distinguish among the mode, median, and mean: The mOde is the value of the mOst frequently occurring score. (For maximum effect, say this aloud, stretching out the long O sounds— mOOOde . . . mOOOst.) The median divides a distribution in half like the median divides a highway in half. And the mean is what we usually *mean* by "average."

3.7   Summing Up Chapter 3

Here is what we have learned in this chapter:

- There are three major kinds of averages: the mode, median, and mean.
- The mode is the most frequently occurring value. It can be calculated for variables measured at any level and is usually the only average appropriate for nominal variables.
- Distributions can be distinguished by the number of modes—unimodal or bimodal.
- The median is the midpoint of an ordered set of scores. Half the scores are less than the median and half are greater. The median can be calculated for ordinal or interval/ratio variables.
- The mean is the arithmetical average. The mean is properly calculated for interval/ratio variables, but can sometimes be used with ordinal variables and dichotomous nominal variables.
- The sum of deviations from the mean $(X_i - \overline{X})$ is 0.
- The mean minimizes the sum of squared deviations of scores $\Sigma(X_i - \overline{X})^2$.
- The mean for a dichotomous variable coded 0 or 1 equals the proportion of cases coded 1.
- The median and mean have the same value in symmetric distributions.
- The mode, median, and mean have the same value in unimodal symmetric distributions.
- Extreme scores (including outliers) affect the mean but not the median or mode.
- The skewness of a distribution affects the mean but not the median. The median is usually preferable to the mean for highly skewed distributions.

Research reports describing averages are usually accompanied by descriptions of standard deviations of variances, the subject of the next chapter. I will wait until we finish Chapter 4, therefore, to offer examples of analysis write-ups for averages.

Key Concepts and Procedures

Ideas and Terms

average

measure of central tendency

mode

unimodal distribution

bimodal distribution

median

mean

deviation

sum of squares

skewness

negatively skewed (left skewed)

positively skewed (right skewed)

Symbols

Mo

Md

$\overline{X}$

μ

X_i

ΣX_i

$\Sigma(X_i - \overline{X})$ and $\Sigma(X_i - \mu)$

$\Sigma(X_i - \overline{X})^2$ and $\Sigma(X_i - \mu)^2$

Formulas

$$Md = L + \left(\frac{\frac{N}{2} - F}{f}\right) \text{(i)}$$

$$\overline{X} = \frac{\Sigma X_i}{N}$$

CHAPTER 4
Measures of Variation

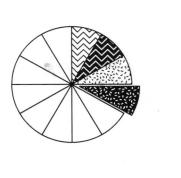

A man drowned crossing a stream with an average depth of six inches. That old tale is enough to remind us that there is far more to attend to about variables than just their averages. Distributions have variations as well as averages, and sometimes variations are far more important than averages. Now we will take up variances and standard deviations—statistics that measure variation. We'll learn to measure how spread out or squished together a variable's distribution is; that is, we will learn to measure how similar or dissimilar scores are from one another. We will also learn about the shapes of distributions and about normal curves. We will use what we know about averages and standard deviations to transform scores into so-called standard or Z-scores that help us compare scores' locations in a distribution. Finally, we will learn to use information about sample averages and standard deviations to estimate averages of a population.

After this chapter you will be able to:

1. Define and calculate standard deviations and variances for both samples and populations.

2. Recognize conditions under which standard deviations and variances are appropriate or inappropriate.

3. Measure skewness in variables.

4. Explain in general terms what a normal curve is and how it may be used.

5. Calculate and interpret standard scores.

6. Explain what a sampling distribution is.

7. Calculate and interpret confidence intervals around the mean.

8. Exercise caution in interpreting averages and measures of variation for aggregate data.

4.1   Variances and Standard Deviations

Now that we know something about averages, let's turn to *measures of variation*[1] that summarize how close together or spread out scores are. For interval/ratio variables, it is often useful to gauge how similar or dissimilar scores are from one another. Are scores similar so that they cluster around some value or are scores spread out quite widely? If scores cluster together, then they are more homogeneous. If scores are spread out widely, then they are more heterogeneous.

Here are three sets of scores that range from relatively similar/homogeneous to relatively dissimilar/heterogeneous:

	Set A Relatively Similar		Set B Somewhere in Between			Set C Relatively Dissimilar		
16	64	−4	44	−24 576	34	−34	1156
	68	0	63	−5 25	58	−10	100
4	70	2	80	12 144	90	22	484
9	71	3	91	23 529	101	33	1089
1	69	1	74	6 36	79	11	121
4	66	−2	56	−12 144	46	−22	484
Mean	68		68		68		

All three sets have the same mean—68. But scores in Set A are relatively similar, hovering closely around the mean of 68, whereas scores in Set B are spread out much more, and scores in Set C range even more widely.

The mean of a set of scores provides a useful reference point for describing variation. We can express variation in terms of how far scores deviate from the mean score. If scores deviate a lot, then the distribution is spread out quite widely from the mean score. If scores do not deviate much, then they cluster about the mean and thus are bunched closely together. We use the mean as a point of reference from which to measure deviations because the mean, unlike the mode or median, takes all scores directly into account. Furthermore, the mean has the useful property that it minimizes the sum of scores' deviations. Remember from our discussion of the mean in Section 3.4 that the sum of deviations of scores from the mean is 0. That's pretty minimal! And recall too that the sum of squared deviations from the mean is a minimum. That is, the sum of the squared differences

---

[1] Measures of variation are sometimes called *measures of variability* or *measures of dispersion*. Terminology varies, but the idea is the same.

between each score and the mean is less than the sum of squared differences between each score and any other number.

The variance and the standard deviation are two very closely related measures of variation that summarize how narrowly or widely scores are distributed around the mean. Let's consider them in turn.

First, the *variance*. I will present two slightly different ways to find variances, one for population data and the other for sample data. The variance of a set of scores for an entire *population* is calculated this way:

1. Calculate the mean.

2. Subtract the mean from each score. (Many of these differences will be negative, but that's OK.)

3. Square each difference. (That is, multiply each difference by itself. Remember that multiplying two negative numbers produces a positive number.)

4. Add all the squared differences.

5. Divide that sum by the total number of scores.

Your answer is the variance of the scores.

I noted in Section 3.4 that the sum of deviations of scores from the mean is always 0. That is, $\Sigma(X_i - \mu) = 0$. That's why in step 3 we square differences between scores and the mean. Otherwise, we would always end up with a sum of 0, which wouldn't tell us anything about variation.[2]

Computing the variance for population data is described very concisely with this formula:

$$\sigma^2 = \frac{\Sigma(X_i - \mu)^2}{N}$$

where σ^2 = variance for population data

X_i = score of the ith case

μ = mean for the population

N = total number of cases in the population

σ is the Greek lowercase sigma; σ^2 is pronounced "sigma squared." A few paragraphs from now I will explain why we use σ^2 to stand for the variance. For now, let me note that we are using a Greek letter because we are referring to the variance of data for a *population* (as opposed to a sample). Recall that μ symbolizes the mean for a popula-

---

[2] We can also base a measure of variation on the sum of the absolute values of the differences, but that measure, called the *mean deviation*, has less useful statistical properties than the variance based on the sum of squares and is not used very often.

tion (rather than a sample). We have already learned that Σ means to add everything that comes after it. In this formula, Σ means to add all the squared differences between each score and the mean. We have seen this sum of squared deviations before (in Section 3.4). It is the sum of squares.

We can't do better than to use the mean as a reference point if we want to minimize the sum of squared deviations. This property of the mean is not surprising since we have already seen (in Section 3.4) that the mean minimizes the sum of simple (i.e., unsquared) deviations (the sum is always 0). Note that by dividing the sum of squares by N, the variance is the average squared deviation of scores from the mean. That's important enough that I want to repeat it: *The variance is the average squared deviation from the mean.*

You can readily understand why the variance is a useful measure of the variation of a distribution. If scores are widely distributed about the mean, then the deviations from the mean are large, and thus the sum of squares is large and the variance is large. And if scores are clustered tightly about the mean, then the deviations, the sum of squares, and thus the variance are all small. Therefore, a larger variance indicates more variation, and a smaller variance indicates less variation.

However, a subtle problem comes up for *sample* data. Although the formula $\sigma^2 = \dfrac{\Sigma(X_i - \mu)^2}{N}$ defines the variance of scores, it is not quite right for sample data. Applied to sample data, the formula gives a biased estimate of the population variance. *Biased* means that if we were to compute the variance of every possible sample of size N drawn randomly from the population, and then average those sample variances, the average sample variance would not quite equal the variance of the population. In fact, the average of all sample variances would always underestimate the population variance. That's a problem—we don't like bias in statistics any more than in social relationships.

Fortunately, statisticians have figured out a way to correct for this bias in calculating the variance for sample data. We divide the sum of squares by N – 1 rather than N. Thus, the estimated population variance based on sample data is given by

$$s^2 = \frac{\Sigma(X_i - \overline{X})^2}{N - 1}$$

where s^2 = variance for sample data

X_i = score of the ith case

$\overline{X}$ = mean of the sample

N = total number of cases in the sample

As a practical matter, whether we use N or N − 1 in the denominator only matters when N is fairly small. In large samples like those of the General Social Survey, for example, it makes little difference whether we divide by 2903 or by 2904. Still, since we so often use sample data like the General Social Survey in social research, I will generally use N − 1 in this text and workbook.

Learning a little terminology here will prove useful later on. The denominator N − 1 is called the *degrees of freedom* of the variance. The reason for this term has to do with the property of the mean noted earlier: The sum of deviations from the mean is always 0. Therefore, if we know all the scores and thus all the deviations but one, we can easily calculate the last score and deviation. The last score is the number that makes the sum of deviations equal to 0. For example, if three scores are 4, 5, and some third score, and the mean of the three scores is 5, then the third score (let's call it X) must be 6. After all:

$$(4 − 5) + (5 − 5) + (X − 5) = 0$$

$$−1 + 0 + X − 5 = 0$$

$$X = 1 + 5$$

$$X = 6$$

In other words, the deviations from the mean are restricted a little. Only N − 1 are free to vary; once they are set, the last one is determined. We say there are N − 1 degrees of freedom. The idea of degrees of freedom will come up often for a variety of statistics throughout this text.

Although we will usually use a computer to find variances, calculating a few by hand helps us understand the variance much better. Here is a simple example of the calculation of the variance. Consider these six scores from a sample: 4, 6, 8, 9, 10, 11. The mean score (calculated in Section 3.3) is 8. To compute the variance, set up a worksheet like this:

X	$X_i - \overline{X}$	$(X_i - \overline{X})^2$
64	64 − 68 = −4	16
68	68 − 68 = 0	0
70	70 − 68 = 2	4
71	71 − 68 = 3	9
69	69 − 68 = 1	1
66	66 − 68 = −2	4
Total	0	34

$$s^2 = \frac{\Sigma(X_i - \overline{X})^2}{N - 1}$$

$$= \frac{34}{6 - 1}$$

$$= \frac{34}{5}$$

$$= 6.80$$

6.80 is the variance of these scores. Note that if you add the column of $X_i - \overline{X}$ differences, the sum is 0. If it isn't, you made a mistake in your calculations.

By the way, here are variances for the three sets of scores we saw a while ago:

	Set A Relatively Similar	Set B Somewhere in Between	Set C Relatively Dissimilar
	64	44	34
	68	63	58
	70	80	90
	71	91	101
	69	74	79
	66	56	46
Mean	68	68	68
s^2	6.80	290.80	686.80

As we surmised from just eye-balling the three distributions, scores in Set A are more alike than scores in Set B, which in turn are more alike than scores in Set C. Now, however, we have quantitative measures of these variations—the variances are 6.80, 290.80, and 686.80. These variances measure how spread out scores are around the mean.

Before moving on, I want to note that for dichotomous variables like gender, coded 0 for male and 1 for female, the variance is $P(1 - P)$, where P is the proportion of cases coded 1 on the variable. In other words, the variance is the proportion of cases coded 1 times the proportion of cases coded 0. We saw in Chapter 3 that the mean of a dichotomous variable coded 0 and 1 is the proportion of cases coded 1. Thus, although we usually apply means and variances to interval/ratio variables, they are also useful in the special cases of dichotomous variables, even nominal ones. I won't bother to prove that the variance of a dichotomous variable is $P(1 - P)$.

A *standard deviation* is the square root of the variance. (That means, of course, that the variance is the square of the standard deviation.) To find the standard deviation of a set of scores:

1. Compute the variance.

2. Find the square root of the variance.

That square root is the standard deviation.

Here are two equivalent formulas for the standard deviation for population data:

$$\sigma = \sqrt{\text{Variance}} \quad \text{and} \quad \sigma = \sqrt{\frac{\Sigma(X_i - \mu)^2}{N}}$$

And here are the comparable formulas for sample data:

$$s = \sqrt{\text{Variance}} \quad \text{and} \quad s = \sqrt{\frac{\Sigma(X_i - \overline{X})^2}{N - 1}}$$

For the sample scores listed above, the standard deviation for Set A is $\sqrt{6.80} = 2.61$. The standard deviations for Sets B and C are $\sqrt{290.80} = 17.05$ and $\sqrt{686.80} = 26.21$.

Reasonably enough, the letter s stands for the standard deviation. That is why the variance, which is the square of the standard deviation, is denoted s^2. Furthermore, just as statisticians use the Roman $\overline{X}$ for a sample mean and the Greek μ for a population mean, so too do they use s for the standard deviation of sample data and σ for the standard deviation of population data. The variance of a population, therefore, is symbolized σ^2.

For scores based on integer data, it is common to round standard deviations and variances to two decimal places, and I will do that throughout this text. There are, of course, exceptions to this general guideline. As always, you need to *think* about the significance of the digits in any particular situation.

Interpretation of variances and standard deviations is straightforward. The formulas show that the less variation there is among scores, the smaller the sum of squares and thus the smaller the variance and standard deviation. If there is no variation at all among scores, the variance and standard deviation are 0. In this case, the scores would all be the same (i.e., all scores would equal the mean) and the "variable" is really a constant. The more variation in scores, the larger the variance and standard deviation. While the lowest possible variance or standard deviation is 0, there is no theoretical upper limit.

Since the variance is just the square of the standard deviation, why have both measures? Why not have *either* the variance *or* the standard deviation? The reason is that the variance and standard deviation each have useful properties. The variance has some mathematical properties that are very important in more advanced statistics. Statistically speaking, statisticians can do a lot with the concept of variance. We will find examples of this when we learn about t tests, analysis of variance, and regression techniques in Chapters 8 through 12.

But the variance has a problem. Because the variance squares deviations from the mean without then "unsquaring" them, it has a different scale than the scores on which it is based. The variance usually doesn't look right because of its different scale. In contrast, the standard deviation usually looks to be in the right scale for a measure of the variation among scores. It should look right, of course, since it is calculated by taking the square root of the average squared deviation from the mean. It "unsquares" a squared number. This unsquaring returns the measure of variation to about the same scale as the original scores.

Consider the three sets of scores from earlier. We have seen that their variances are 6.80, 290.80, and 686.80. Each "runs" larger—much larger—than the set of scores it describes. Even the smallest—6.80—looks too large as a measure of variation for scores that range from 64 to 71. But here are the standard deviations of Sets A, B, and C: 2.61, 17.05, and 26.21. They seem in the same scale as the scores on which they are based.

Or consider the ages of respondents in the General Social Survey. Scores range from 18 to 89, with a mean of 44.8. The variance of age is a huge-looking 284.5, but its standard deviation is a reasonable-looking 16.9, in the same scale of years as respondents' ages. In reporting results of statistical analyses, therefore, we usually report the standard deviation rather than variance because the standard deviation makes more cognitive sense to our human minds. For the most part, then, use the variance for more advanced statistical reasoning and rely on the standard deviation in reporting results.

Since the variance and standard deviation are based on deviations from the mean, and since (strictly speaking) the mean can be calculated only for interval/ratio variables, the variance and standard deviation are calculated appropriately only for variables measured at the interval/ratio level. Strictly speaking, do not calculate the variance or standard deviation of nominal- or ordinal-level variables. But as we found for averages, sometimes it is useful to break this rule and calculate variances and standard deviations for dichotomous nominal variables and for ordinal variables if they tell us useful information about the variable.

Watch out for extreme scores, especially outliers, that can unduly affect the variance and standard deviation. We have already seen that unusually low or unusually high scores can make the mean unrepresentative. The effect of extreme scores is even greater on measures of variation. The variance and standard deviation not only use deviations from the mean, they *square* those deviations, greatly increasing their impact. Squaring numbers larger than 1 increases them exponentially—that is, increases them a lot. The deviation of a score of 4 from a mean of 8 is only 4, but the *square* of that deviation is a whopping 16.

In extreme situations, a variance or standard deviation may be inflated entirely because of a single outlier, especially if the number of cases N is small. Here's an example: For the percentage of a state's population of Asian ancestry, the standard deviation is 8.65 with Hawaii included and only 1.54 with Hawaii excluded. I pointed out in Section 3.4 that Hawaii is an extreme outlier with 61.8 percent Asian, and its score has an undue effect on the standard deviation. In such situations you should first consider why these extreme scores occur. Then consider excluding them from your analysis when calculating the mean, variance, and standard deviation. Also remember to exclude missing data when computing variances and standard deviations.

4.2 Shapes of Distributions

Like people and dogs, distributions come in all kinds of shapes. We have already seen that some distributions are unimodal, some are bimodal, and some are too flat to be described in terms of modes. Some distributions have long tails to the right, some have long tails to the left, some have both, and some have no tails at all. Some distributions are tall and skinny, some are short and wide, some are in between. Let's take a brief look at shapes of distributions.

Statisticians use the term *kurtosis* to refer to the peakedness of distributions of interval/ratio variables. *Leptokurtic* distributions are tall and thin; *platykurtic* distributions are low and flat; and *mesokurtic* distributions are in between. Figure 4.1 shows examples of these three general shapes of distributions.

Figure 4.1. General Shapes of Distributions

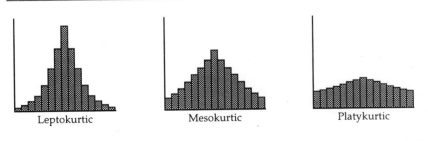

Leptokurtic   Mesokurtic   Platykurtic

Skewness is also an important characteristic of distributions. As we saw in Section 3.6, some variables are distributed symmetrically whereas others are skewed to the right or left. Statisticians have devised a measure of skewness based on the different effects of skewness on the mean and median. Remember from Section 3.5 that extreme scores pull the mean toward them while the median depends only on the middle score in a distribution, so the mean is always further in the direction of the skew. The following formula makes use of these properties of the mean and median to measure skewness:

$$\text{Skewness} = \frac{3(\overline{X} - Md)}{s}$$

The formula calls for dividing by the standard deviation in order to adjust for the variable's scale. (Why we multiply $(\overline{X} - Md)$ by 3 is more complicated and not worth considering here.)

Notice in this formula that the difference between the mean and median gauges the extent of skewness. The more skewed a distribution, the greater the difference between mean and median, and the larger the numerator. Skewness is positive for distributions skewed to the right and negative for distributions skewed to the left. You can see why by comparing the relative positions of the means and medians in the left-skewed and right-skewed diagrams in Figure 3.5. Skewness is 0 for symmetric distributions because then the mean equals the median, and the numerator is 0.

An example: Here are the means, medians, and standard deviations for the years of education of males and females in the General Social Survey:

Statistic	Males	Females
Mean	13.56	13.21
Median	13.12	12.66
Standard Deviation	2.95	2.91

And here are the skewness coefficients for these two distributions:

Males	Females
$\text{Skewness} = \dfrac{3(\overline{X} - Md)}{s}$	$\text{Skewness} = \dfrac{3(\overline{X} - Md)}{s}$
$= \dfrac{3(13.56 - 13.12)}{2.95}$	$= \dfrac{3(13.21 - 12.66)}{2.91}$
$= \dfrac{3(.44)}{2.95}$	$= \dfrac{3(.55)}{2.91}$

$$= \frac{1.32}{2.95} \qquad\qquad = \frac{1.65}{2.91}$$

$$= .45 \qquad\qquad\qquad = .57$$

The skewness coefficients indicate that the education distributions of both males and females are positively skewed, with the education distribution of females skewed considerably more than that of males (.57 for females versus .45 for males).

A caution: Do not assess skewness only by comparing means and medians or by calculating skewness coefficients. Mean-median comparisons and skewness coefficients can be deceiving for oddly shaped distributions. Be sure to visually inspect the shape of a distribution by looking at its graph. We would, for example, ordinarily consider the graphs of distributions along with the education skewness coefficients calculated above.

4.3 Standard Scores (Z-Scores)

A *standard score*—often called a *Z-score*[3]—describes how many standard deviations from the mean a score is located. Z-scores are particularly useful when we compare scores from distributions that have different means and standard deviations. For example, which of the following Yale University law students has a better score on his/her test, relative to their respective classes: Bill, with a score of 87 on a test with a class mean of 81 and a standard deviation of 6; or Hillary, with a score of 83 on a test with a class mean of 76 and a standard deviation of 4? We can answer this question by converting test scores to standard scores and then comparing standard scores.

To convert a score into a Z-score:

1. Subtract the mean from the score.

2. Divide that difference by the standard deviation.

---

[3] More precisely, Z-scores are transformations of scores only on normally distributed variables, whereas standard scores are transformations of *any* variable, even ones not normally distributed. The terms *Z-score* and *standard score* are often used interchangeably, however, and that is what I will do in this text.

As a formula:

$$Z_i = \frac{X_i - \overline{X}}{s}$$

where Z_i = standard score of the ith case

X_i = score of the ith case

$\overline{X}$ = mean

s = standard deviation

Obviously Z-scores may be either positive or negative since scores may be either greater than or less than the mean. A positive sign indicates that the score is above the mean; a negative sign indicates that the score is below the mean.

Here are the Z-scores for Bill's and Hillary's exam scores, 87 and 83, offered above:

<u>Bill's Score</u>   <u>Hillary's Score</u>

$$Z_i = \frac{X_i - \overline{X}}{s} \qquad\qquad Z_i = \frac{X_i - \overline{X}}{s}$$

$$= \frac{87 - 81}{6} \qquad\qquad = \frac{83 - 76}{4}$$

$$= \frac{6}{6} \qquad\qquad\qquad = \frac{7}{4}$$

$$= 1.00 \qquad\qquad\qquad = 1.75$$

Bill's test score of 87 lies a respectable 1.00 standard deviation above the mean score for his class, but Hillary's score of 83 is a more impressive 1.75 standard deviations above the mean of her class. Relative to their respective class distributions, Hillary's score of 83 is better than Bill's score of 87. Her score is further above the mean if distance is measured in standard deviation units.

A *standardized variable* is one whose scores have all been converted to standard scores. We say that such scores are in *standard form*. That is, each score has been transformed into the number of standard deviations it lies from the mean. All standardized variables have the same scale, one with a mean of 0 and a standard deviation of 1.00. That is, any distribution of Z-scores has a mean of 0 and standard deviation of 1.00. Z-scores do not change the relative position of scores. High scores are still relatively high; low scores are still relatively low.

Here is the calculation of Z-scores for the six cases ($\overline{X}$ = 68 and s = 2.61) considered earlier in this chapter:

X_i	$X_i - \overline{X}$	$(X_i - \overline{X})/s$	=	Z_i
64	64 – 68 = –4	–4/2.61	=	–1.532
68	68 – 68 = 0	–0/2.61	=	.000
70	70 – 68 = 2	2/2.61	=	.766
71	71 – 68 = 3	3/2.61	=	1.149
69	69 – 68 = 1	1/2.61	=	.383
66	66 – 68 = –2	–2/2.61	=	–.766
Total	0.000			

Note that the sum of these scores in standard form is 0. Note, too, that since $\frac{\Sigma Z_i}{N} = \frac{0}{6} = 0$, these standardized scores' mean is also 0. I'll let you calculate the standard deviation of these Z-scores if you wish. I assure you that the standard deviation is 1.00. Every distribution of Z-scores has a mean of 0 and standard deviation of 1.00.

Z-scores have another interesting characteristic. The sum of the squares of Z-scores is equal to N, the number of cases. That is, ΣZ^2 = N, where Z denotes standardized scores. (I'll let you demonstrate this with the six Z-scores calculated above.) We will make use of this property of Z-scores when we learn about correlation coefficients in Chapter 10.

A warning (an obvious one) about Z-scores: Because they are based on the mean and standard deviation, Z-scores are meaningful only for interval/ratio variables. Do not calculate or use Z-scores for nominal or ordinal variables.

Another warning about Z-scores: Converting scores to standard scores does *not* transform a non-normal distribution into a normal distribution. If a variable is not distributed normally, the distribution of its standard scores will also not be normal.

4.4 Normal Distributions

One particular kind of distribution—the *normal distribution*—is so important that we need to give it special attention. You may already know that a normal distribution is symmetric and bell-shaped. You may not know, however, that although there are many normal distributions, not every symmetric, bell-shaped distribution is normal—only those that fit the following complicated formula:

$$Y = \frac{e^{-(X-\mu)^2/2\sigma^2}}{\sigma\sqrt{2\pi}}$$

Fortunately, we can learn what we need to know about normal distributions without using this formula, so we don't need to know much about it. (You will learn more about this formula in an advanced statistics course.) I want you to see the formula, however, so that I can point out that normal distributions depend only on the mean μ and the standard deviation σ. (Note that I am using notation for populations rather than samples here.) The other terms, e (the base of natural logarithms, or 2.71828...) and π (pi, or 3.14159...), are constants. There are, in fact, an infinite number of normal distributions, one for each combination of possible means and standard deviations.

A normal distribution with a mean of μ and a standard deviation of σ is denoted $N(\mu, \sigma)$. If a set of scores has a distribution $N(85, 12)$, then it has a normal distribution with a mean of 85 and a standard deviation of 12. For a given mean (μ), a normal distribution may be tall and thin (if σ is small) or short and flat (if σ is large).

Figure 4.2 shows examples of two distributions, each normal but with different standard deviations. The normal distribution on the left has a larger standard deviation, of course, than the distribution on the right. The value of the mean μ shifts these distributions to the left (for a smaller μ) or right (for a larger μ). Since all normal distributions are symmetric, the mean is always at the center of a distribution and always equals the median and mode. Regardless of the mean and standard deviation, the points of inflection where the normal curve changes shape from concave to convex are always 1 standard deviation from the mean.

Figure 4.2. Normal Distributions with Different Standard Deviations

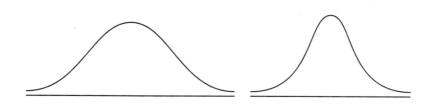

But while normal distributions are not all the same, they share an important characteristic: A given standard deviation from the mean always "cuts off" the same proportion or percentage of scores in all normal distributions. So, for example, in a normal distribution, 1 standard deviation from the mean includes about 68 percent of scores, 2 standard deviations include a little over 95 percent, and 3 standard deviations from the mean include 99.7 percent of scores. Figure 4.3

displays these whole-number standard deviations and their respective percentages. These percentages come up often enough that you should memorize them: 68–95–99.7.

Figure 4.3. Areas Under a Normal Curve

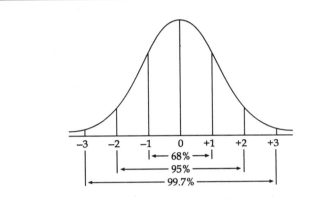

Sometimes it is more useful to begin with a given percentage of cases and then describe with a little more precision how many standard deviations from the mean embrace that percentage. For example, in a normal distribution 95 percent of scores lie within 1.96 standard deviations of the mean, and 99 percent lie within 2.58 standard deviations. These particular distances from the mean—1.96 and 2.58 standard deviations—will come up again when we consider confidence intervals in Section 4.6.

Despite its name, a normal distribution is no more "right" or "proper" than any other distribution although a surprising number of variables in the natural world have distributions that are approximately normal. The number of kernels on ears of corn, the number of hairs on men's or women's heads, the number of ants in anthills, the number of stars in galaxies . . . and on and on. Distributions on many standardized tests are also normal, although that is not really surprising since standardized tests often are scored in a way that produces a normal distribution. Alas, normal distributions are much rarer in the social than the biological or physical worlds. Many of our variables like income or education are quite skewed and thus far from normal.

Nevertheless, the normal distribution plays a central role in statistical reasoning, as we will find several times throughout this text. Even for variables that are not themselves normally distributed, certain important statistics based on the variables are almost normally distributed. We will use this important fact in the next section when we take up methods of generalizing from sample data to the populations from which they are drawn.

4.5   Sampling Distributions

We can use distributions of sample data to describe the population from which the sample was drawn. First, however, we need to learn what a sampling distribution is. An example will help. Consider the variable IQ measured by scores on a standardized IQ test. I pick IQ as an example only because it is a particularly convenient one, not because I "believe" in IQ tests. Unlike almost all other variables we work with in sample data, IQ scores have a known population distribution. Many standardized IQ tests are scored in such a way that if we were to give the test to a large population like all adult Americans, the distribution of scores forms a normal distribution like those we saw in the previous section. Moreover, the mean score for this large population is 100. In symbols: $\mu = 100$.

Now carry out what the Germans call a gedanken experiment—a thought experiment using our imaginations. Suppose we were to administer an IQ test to all 200,000,000 adult Americans. That would not be practical, of course. In fact it would be virtually impossible. But suppose we did this gigantic study. As I noted in the last paragraph, what we would find is that these 200,000,000 scores are normally distributed with a mean of 100.

Now suppose we do something that certainly would be possible. Suppose we select a random sample of, say, 1500 adult Americans and measure each of their IQs. How would these 1500 scores be distributed? What would the sample mean $\overline{X}$ be? Answer: We don't know for sure until we look at the actual distribution of the 1500 scores of those sampled.

It is tempting—but wrong—to think that a random sample of 1500 cases must itself be normally distributed with a mean of 100 just like the population from which the sample is randomly drawn. Although the sample *could* have a normal distribution with $\overline{X} = 100$, it probably wouldn't, at least not exactly. After all, since the sample is random, it *might* happen—just by chance—that the sample consists of 1500 adults with IQs a little above the population average of 100. Or maybe, just by chance, we might end up with a sample whose mean IQ is a little below average. We could even end up with a sample with the 1500 highest IQs in the country (and thus a very high $\overline{X}$). That's extraordinarily unlikely, but it *could* happen. And likewise, the sample *could* consist of Forrest Gump and the other 1499 least intelligent adults (and thus have a very low $\overline{X}$). That too is extraordinarily unlikely, but it too *could* happen. And there are a zillion other possible samples too, each with its own IQ distribution and its own IQ mean. Most samples will be pretty much like the larger population from which they are drawn, but some samples will be somewhat different from the population and some will even be grossly unrepresentative of the population. A sample *could* consist entirely of nuns, for example, or entirely of motorcyclists, or even entirely of motorcycle-riding nuns. Yes, there are lots of possible samples that *could* happen. Fortunately, though, most of the possible samples we could draw will be pretty much like the population.

Since the sample we are imagining for IQ scores is random, there is no telling what its distribution will be until we analyze the scores in the sample. Sure, most possible samples will probably be shaped pretty much like the population, with means around the population mean of 100. But some samples will have much higher means and some will have much lower means. Some samples will even be quite different from the population. But let's not let these relatively few extreme samples distract us from the important fact that most possible samples will have means somewhere around the population mean.

Now suppose we do something that would be totally impossible (remember, this is a thought experiment). Suppose we select every possible random sample of 1500 people that could possibly be drawn. That's a zillion different samples. *You* would be in some of those samples (although not in most of them). I'd be in some samples too (although, again, not in most). You and I would even be in some samples together. We would even be together with Madonna and Michael Jackson in some samples (a scary thought).

Suppose, further, that we calculate the mean for each of these zillion possible samples. And then suppose we plot the distribution of all these sample means. What we would have is a sampling distribution. More generally, a *sampling distribution* is a distribution of statistics (in this example, means) of all possible samples of a given size (here, 1500 cases) drawn from some population (like all 200 million American adults). A sampling distribution of the mean describes the number or the proportion of samples with means of various sizes.

Note that there are three kinds of distributions that we need to distinguish:

Population Distribution — The distribution of scores in a population.

Distribution of a Sample — The distribution of scores in a sample of a given size.

Sampling Distribution — The distribution of some statistic (e.g., the mean) in all possible samples of a given size.

The last two distributions have similar sounding names, but they are quite different. Figure 4.4 presents a schematic depiction of these three distributions.

Even though we cannot actually draw all possible samples of a given size from a population as large as 200,000,000 (there are far too many possible samples), statisticians have used mathematics to figure out what the sampling distribution is for certain important statistics. If we make some assumptions about the population and if we draw a sample randomly, then we know how some statistics (like the mean, for example) are distributed for all possible samples of a given size that could be drawn. In other words, we know their sampling distributions.

The sampling distribution of the mean for a random sample has extremely important properties. As the sample size N increases, the sampling distribution of the mean more and more closely resembles a normal distribution with a mean equal to the population mean and a standard deviation of $\frac{\sigma}{\sqrt{N}}$. We can describe this distribution symbolically as $N(\mu, \frac{\sigma}{\sqrt{N}})$. Statisticians refer to this tendency as the *central limit theorem,* one of the most important ideas in statistics.

Figure 4.4. Population Distribution, Distributions of Samples, and Sampling Distribution

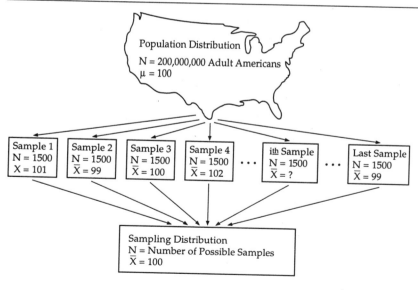

In fact, the sampling distribution of the mean approximates a normal distribution fairly closely for sample sizes of 30 or more. This is true regardless of the shape of a variable's distribution in the

population. Thus, even if a variable is not normally distributed in the population, the mean of all possible sample means is the same as the population mean, and the standard deviation of the distribution of all possible sample means is $\frac{\sigma}{\sqrt{N}}$.

Remember from the previous section that in a normal distribution a given proportion of cases lie within a specified number of standard deviations from the mean. OK, now let's put that fact together with the central limit theorem. The central limit theorem tells us that we can find the proportion of "cases" (i.e., sample means) that lie a given number of standard deviations from the mean of the sampling distribution. We will do that in the next section.

But first some terms and formulas. The standard deviation of a sampling distribution is so important that it has a special name and symbol—*standard error*, designated $\sigma_{\overline{x}}$. As we have just seen, the standard error of the mean—i.e., the standard deviation of the sampling distribution of means from all possible samples of a given size drawn randomly from some population—is given by the formula:

$$\sigma_{\overline{x}} = \frac{\sigma}{\sqrt{N}}$$

For example, the variable daily hours of television watching has a standard deviation of 2.14 based on 1940 cases in the GSS. Using s to estimate σ, we find the standard error of daily television watching:

$$\sigma_{\overline{x}} = \frac{\sigma}{\sqrt{N}}$$

$$= \frac{2.14}{\sqrt{1940}}$$

$$= \frac{2.14}{44.045}$$

$$= .049$$

The standard error—the standard deviation of the sampling distribution—is .049. We will make good use of this standard error in the next section. I have carried calculations out to three decimal places because the standard error is used in calculations of other statistics in the next section.

4.6   Confidence Intervals

Our best estimate of the population mean is the sample mean, but we know that estimate is unlikely to be exactly right because the sample was selected randomly and random samples vary. The population mean may be greater or may be less than the sample mean. It would be very useful, therefore, to have a range around the sample mean within which we think the population mean probably lies. This range is called a *confidence interval*.

The standard error allows us to find a confidence interval. From our IQ thought experiment in the previous section, we know the population mean for IQ scores: $\mu = 100$. But that knowledge is unusual. More typically in research, we do not know what the population mean is and we use our sample mean to estimate it. In technical terms, we use a sample statistic to estimate a population parameter. After all, if we already knew the population mean, we would have little interest in sample data.

The central limit theorem tells us that as N increases in size, the sampling distribution of the mean more and more closely resembles a normal distribution with a mean equal to the population mean μ and a standard deviation $\dfrac{\sigma}{\sqrt{N}}$ (i.e., $\sigma_{\overline{X}}$). Recall from our discussion of normal distributions in Section 4.4 that 95 percent of scores lie within 1.96 standard deviations of the mean. We can use this knowledge to find a confidence interval. To find the **95 percent confidence interval**, simply subtract 1.96 standard errors from the sample mean to find the lower limit and add 1.96 standard errors to the sample mean to find the upper limit of the confidence interval. Expressed as a formula:

$$\text{95 percent confidence interval} = \overline{X} \pm 1.96\sigma_{\overline{X}}$$

The chances are 95 out of 100 that the population mean lies within this range.

Researchers often use a broader **99 percent confidence interval**. Recall from Section 4.4 that 99 percent of scores lie within 2.58 standard deviations of the mean. Thus, we find the 99 percent confidence interval with this formula:

$$\text{99 percent confidence interval} = \overline{X} \pm 2.58\sigma_{\overline{X}}$$

The chances are 99 out of 100 that the population mean lies within this interval.

As an example of confidence intervals, take a variable such as hours per day watching television. In the General Social Survey, this variable has a mean of 2.90, with a standard deviation of 2.14. We found the standard error of the mean in the last section: $\sigma_{\overline{X}} = .049$.

To find the 95 percent confidence interval:

$$\text{95 percent confidence interval} = \overline{X} \pm 1.96\sigma_{\overline{X}}$$
$$= 2.90 \pm 1.96(.049)$$
$$= 2.90 \pm .096$$
$$= 2.80 \text{ to } 3.00$$

Thus the 95 percent confidence interval is between 2.80 and 3.00. The chances are 95 out of 100 that the population mean lies within this range.

Similarly, the 99 percent confidence interval:

$$\text{99 percent confidence interval} = \overline{X} \pm 2.58\sigma_{\overline{X}}$$
$$= 2.90 \pm 2.58(.049)$$
$$= 2.90 \pm .126$$
$$= 2.77 \text{ to } 3.03$$

There is a 99 percent chance that the population mean lies between 2.77 and 3.03.

Incidentally, you are now doing inferential statistics. You are using sample data to infer characteristics of the population from which the sample was drawn.

4.7 Some Cautions Using Univariate Statistics

Finding averages, variances, standard deviations, Z-scores, and confidence intervals is simple, but there are three data-related problems: (1) inappropriate levels of measurement, (2) categories of unequal width (including open-ended categories), and (3) missing data. This section describes each of these three problems.

First, statistical software like MicroCase does not take level of measurement into consideration in generating statistics. The mean, standard deviation, and variance usually are inappropriate for nominal and ordinal variables, and the median is unsuitable for nominal variables. Confidence intervals for means are usually not appropriate for nominal or ordinal variables because the mean itself is usually inappropriate. But computers will gladly compute whatever they are programmed to compute even if what they are programmed to compute is inappropriate—like means, standard deviations, and variances for ordinal variables, and these statistics plus medians for nominal variables.

The point is important: We must be very cautious interpreting computer output. In particular, we must be careful to attend only to those averages, standard deviations, variances, and confidence intervals for means that are meaningful given the level of measurement of our variables. Usually these statistics require interval/ratio levels of measurement.

As always, the most important process in doing statistical analysis is *thinking*. Don't let the computer do your thinking for you.

A second problem arises in finding means, standard deviations, and variances for variables with categories of unequal widths. Consider, for example, Table 4.1 showing the frequency distribution of the number of children of GSS respondents. Note that the coding of the variable combines all 25 cases with more than 8 or more children, coding each score as 8. This open-ended category may well contain respondents with 9 kids, or 10, or 11, or even more. Because such higher scores on this variable are counted only as 8 each, putting them all in the same category has the effect of reducing the mean and reducing the variance and standard deviation. Granted, because the number of cases with 8 or more children is rather small (only 25, or .9 percent of the total), the effect in this sample is probably small. But we should be aware of even these effects, and be very careful of them when the proportions of cases in open-ended categories are very large.

Table 4.1. Number of
Children
(in frequencies)

Number of Children	f
0	882
1	461
2	770
3	420
4	222
5	94
6	48
7	27
8 or more	25
(N)	(2889)

This general problem occurs not just with open-ended categories, but whenever a variable has categories embracing a range of values. This most typically happens with collapsed categories or an open-ended category at the end of a distribution (as with number of children in the example above). Although there is no simple solution to this problem, it must be taken into consideration when we interpret means and measures of variation. Again, *thinking* (yours, not the computer's) is crucial in statistical analysis. Think about the variables you are analyzing and the implications of their measurement and values for your statistical findings.

A final problem: missing data. As a general rule, exclude cases with missing data when calculating averages, measures of variation, and confidence intervals, just as we exclude missing data when calculating percentages. Inclusion of values like Don't Know and No Answer distorts all statistics. Typically coded with high numbers (7, 8, or 9 for one-digit codes and 97, 98, or 99 for two-digit codes), inclusion of such values would improperly increase the mean, standard deviation, and variance.

4.8   Summing Up Chapter 4

Here is what we have learned in this chapter:

- The variance and standard deviation measure how widely scores are spread out around the mean. They can be calculated for interval variables.

- The variance of a dichotomous variable coded 0 and 1 is the proportion of cases coded 0 times the proportion of cases coded 1.

- Extreme scores can have large effects on the variance and standard deviation.

- Scores for interval/ratio variables can be converted into standard or Z-scores by subtracting the mean and dividing by the standard deviation. Standard scores allow us to compare the relative locations of cases in distributions.

- A standardized variable is one whose scores have been transformed into standard scores.

- Standardized variables have a mean of 0 and a standard deviation of 1.00.

- The sum of squares of standardized scores is N.

- A normal distribution is symmetric and bell-shaped, but not all symmetric and bell-shaped distributions are normal.

- In all normal distributions, a given standard deviation from the mean always includes the same proportion of scores.

- A sampling distribution is the distribution of a sample statistic (e.g., mean) for all possible samples of a given size drawn from a particular population.

- A standard error is the standard deviation of a sampling distribution.
- A confidence interval describes the chances that a population parameter such as the mean lies within a specified range.
- Missing data should be excluded from analyses.
- We must carefully interpret computer-generated averages, measures of variation, and confidence intervals for means, guarding against statistics inappropriate for a variable's level of measurement and recognizing the effects of open-ended categories and categories of unequal width.

Key Concepts and Procedures

Ideas and Terms

measures of variation
variance
deviation from the mean
degrees of freedom
standard deviation
kurtosis
leptokurtic
platykurtic
mesokurtic
skewness

standard score
Z-score
standardized variable
standard form
normal distribution
sampling distribution
Central Limit Theorem
standard error
confidence interval

Symbols

σ and σ^2
s and s^2
P and $(1 - P)$
Z
$\sigma_{\overline{X}}$

Formulas

$$\sigma^2 = \frac{\Sigma(X_i - \mu)^2}{N}$$

$$\text{Skewness} = \frac{3(\overline{X} - Md)}{s}$$

$$s^2 = \frac{\Sigma(X_i - \overline{X})^2}{N - 1}$$

$$Z_i = \frac{X_i - \overline{X}}{s}$$

$$\sigma = \sqrt{\text{Variance}} \qquad\qquad\qquad \sigma_{\overline{X}} = \frac{\sigma}{\sqrt{N}}$$

and

$$\sigma = \sqrt{\frac{\Sigma(X_i - \mu)^2}{N}}$$

95 percent confidence interval $= \overline{X} \pm 1.96\sigma_{\overline{X}}$

99 percent confidence interval $= \overline{X} \pm 2.58\sigma_{\overline{X}}$

$$s = \sqrt{\text{Variance}}$$

and

$$s = \sqrt{\frac{\Sigma(X_i - \overline{X})^2}{N - 1}}$$

Analysis Write-up 2: Averages and Standard Deviations

✦   ✦   ✦   ✦   ✦

We can often report either medians or means and standard deviations of variables right in the text of an analysis write-up without using a table. Tables, however, are more efficient for presenting means and standard deviations when we need to describe numerous variables.

Table 1 presents education and occupational prestige averages and standard deviations for males in the General Social Survey along with comparable statistics for their parents. Both medians and means for education register a substantial increase in average education. Male respondents have a median education over a year greater than their parents, and mean education has increased about two years. However, variation in education, assessed with standard deviations, has decreased considerably, especially when compared with the variation in father's education. Respondents report occupations averaging only slightly less prestige than their fathers, although the variation in prestige scores is much greater for respondents. The slightly lower prestige scores of GSS respondents compared with their fathers may reflect differences in age and life-cycle stage.

TABLE 1 ABOUT HERE

Include the table at the end of the paper or report:

Table 1. Medians, Means, and Standard Deviations for
Education and Occupational Prestige (Males Only)

Variable	Median	Mean	SD	(N)
Years of Education				
Respondent	13.1	13.6	2.95	(1283)
Father	12.0	11.7	3.99	(964)
Mother	12.0	11.6	3.28	(1109)
Occupational Prestige				
Respondent	41.7	43.4	13.87	(1267)
Father	42.4	44.3	12.50	(1069)
Mother	40.2	41.0	13.37	(761)

In this example, I used the subset command to restrict the analysis to males. More refined analyses would incorporate the variable age, an especially important variable when comparing occupational prestige of respondents and their parents. Respondents may well be at a different age and thus career stage than their parents were. More refined analysis, too, would divide standard deviations by means. This measure adjusts for tendencies toward larger standard deviations for variables with higher means. In this case, however, means are quite similar, so dividing standard deviations by them would have little effect on comparisons of the variability of variables.

PART II
Bivariate Analyses

CHAPTER 5
Cross-tabulation

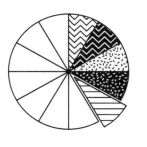

Cross-tabulation (or *cross-tab* for short, and sometimes called *cross-classification* or *tabular analysis*) examines the association between two variables by comparing percentage distributions. Sometimes we are interested simply in whether two variables are associated with one another without concern about causality. For example, we might want to know if reading ability and mathematical ability are associated with one another without treating one as caused by the other.

More typically, however, we are interested in causal relationships in which we hypothesize that an independent variable affects a dependent variable. We might hypothesize, for example, that amount of education (the independent variable) influences attitude toward civil disobedience (the dependent variable), or that education affects amount of television watching, or that gender affects fear of walking in one's neighborhood. In this chapter and the next two chapters we will learn to create, interpret, and present bivariate tables to study these sorts of relationships.

Cross-tabulation is only one of several methods to analyze bivariate relationships. We will consider some other important methods—difference of means, analysis of variance, and regression and correlational techniques—in Chapters 8, 9, and 10. In Part III we will extend these statistical procedures by introducing additional variables—multivariate analysis. But first, cross-tabulation.

All these statistical methods—tabular analysis, analysis of variance, and regression and correlation—offer ways to answer six key questions about a relationship between two variables:

a. Is there a relationship between the two variables in the data we are analyzing?

b. How strong is the relationship?

c. What is the direction and shape of the relationship?

111

d. How confident can we be that the relationship is not due to random processes, and if data are from a sample, can we generalize the relationship to the population from which the sample was drawn?

e. Is the relationship genuinely causal, or is it instead a spurious relationship caused by some third variable?

f. What intervening variable(s) link or connect the independent and dependent variables?

These questions will concern us over and over through the rest of this text. We will learn some ways to answer the first three questions in this chapter and some other ways in Chapters 8 through 10. Question d concerns generalizing sample results to a population—we will take that up in Chapter 6 and again in Chapters 8 through 11. Questions e and f involve multivariate analysis; we will address them in Chapters 11 and 12. By the time we're done, we will know quite a bit about bivariate and multivariate analyses of relationships among variables.

After this chapter you will be able to:

1. Transform sets of raw data into bivariate frequency tables.
2. Transform bivariate frequency tables into percentage tables.
3. Interpret bivariate percentage tables.
4. Present properly formatted bivariate tables.
5. Identify positive, negative, and curvilinear relationships in bivariate percentage tables.
6. Construct and interpret stacked bar graphs.
7. Recognize problems caused by small Ns in bivariate tables and address such problems by collapsing variables or excluding values.
8. Understand that association does not imply causation.

5.1   Bivariate Frequency Tables

A cross-tabulation of two variables requires a statistical table with columns and rows. The values of one variable label the columns; the values of the other variable label the rows. Construction of a bivariate table is analogous to the construction of a univariate table described in Sections 2.1 and 2.2. As an example, consider the relationship between education and civil disobedience, two variables we examined in univariate tables in Chapter 2. Education may well foster reliance on internalized values rather than external authority, encouraging following the dictates of conscience when it conflicts with the law. We might hypothesize, therefore, that the amount of education a person completes affects his/her opinion about whether people should always obey the law or instead should follow their consciences even if it means breaking the law.

Education in this example is the *independent variable*; attitude toward civil disobedience is the *dependent variable*. An independent variable is the presumed or hypothesized *cause* of the dependent variable. Saying the same thing another way: The dependent variable is the presumed or hypothesized *effect* of the independent variable. Sometimes we represent a causal relationship with an arrow going from the independent variable to the dependent variable, like this: Education → Civil Disobedience. Of course, we don't know for sure that the variable Civil Disobedience is associated with Education. This is a *research hypothesis*—an expectation that these two variables are related. We can test this Education → Civil Disobedience hypothesis by analyzing data using cross-tabulation techniques.

Consider a sample of 50 cases for whom we have information on education and civil disobedience. Table 5.1 reports the Education and Civil Disobedience scores of this subsample of the General Social Survey. We met these same 50 cases in Section 2.1, although here I have collapsed the Junior College, B.A., and Graduate Degree values into a single College category. We will find as we learn bi- and multivariate techniques that we can sometimes learn more about a relationship by collapsing categories even though collapsing loses some detailed information. Besides, dealing with three rather than five categories of the independent variable will keep our cross-tabulation example simpler.

If you tally the univariate distributions of education and civil disobedience, you will find they are the same as the univariate examples in Section 2.1 (allowing, of course, for collapsed values of Education). Now, however, we are interested in whether these variables are associated with one another. Are people with more education more likely to believe in following one's conscience even if it means breaking the law? A bivariate table can answer this question.

A *bivariate frequency distribution* for these data is set up as in Table 5.2, with categories of education across the top and civil disobedience down the stub, or side, of the table. To create a bivariate frequency table like this, count the number of cases that have each combination of scores on the two variables and enter that number in the corresponding cell of the table. As noted above, this process is directly analogous to constructing frequency tables for single variables, although here I have skipped showing you the intermediate step of tallying frequencies.

Table 5.1. Education and Civil Disobedience (50 Cases)

Case Number	Education	Civil Disobedience	Case Number	Education	Civil Disobedience
01	High School	Conscience	26	Grad Degree	Conscience
02	College	Conscience	27	< High School	Conscience
03	< High School	Obey Law	28	College	Conscience
04	High School	Obey Law	29	< High School	Obey Law
05	High School	Conscience	30	High School	Obey Law
06	High School	Obey Law	31	< High School	Obey Law
07	College	Obey Law	32	High School	Conscience
08	High School	Conscience	33	High School	Obey Law
09	High School	Conscience	34	Junior College	Obey Law
10	High School	Conscience	35	High School	Conscience
11	Grad Degree	Conscience	36	College	Conscience
12	High School	Obey Law	37	Grad Degree	Conscience
13	High School	Obey Law	38	High School	Obey Law
14	High School	Obey Law	39	< High School	Obey Law
15	< High School	Obey Law	40	High School	Conscience
16	High School	Conscience	41	College	Conscience
17	High School	Conscience	42	High School	Conscience
18	Junior College	Conscience	43	High School	Conscience
19	Grad Degree	Obey Law	44	< High School	Conscience
20	< High School	Obey Law	45	High School	Obey Law
21	High School	Obey Law	46	< High School	Conscience
22	College	Conscience	47	College	Obey Law
23	Junior College	Conscience	48	College	Conscience
24	High School	Conscience	49	High School	Obey Law
25	College	Conscience	50	< High School	Conscience

The numbers in the table are called *joint* or *cell frequencies*. Each cell frequency reports the number of cases that have that combination of values of row and column variables. For example, 4 respondents (Cases 27, 44, 46, and 50) are high school dropouts who endorse following one's conscience, 6 respondents (Cases 3, 15, 20, 29, 31, and 39) with less than a high school education believe in always obeying the law, and so on.

Table 5.2. Civil Disobedience by Education (in frequencies)

| Civil Disobedience | Education | | | |
	Less Than Hi School	High School	College	Total
Conscience	4	13	12	29
Obey Law	6	11	4	21
Total	10	24	16	50

Note that each data column is the univariate distribution of Civil Disobedience for cases with that particular value of Education. The first column, for example, is the univariate distribution of Civil Disobedience for respondents with less than a high school education.

The row totals and column totals are called *marginal distributions*, or just *marginals* for short. They present the univariate distribution of each variable. As I mentioned earlier, the marginal distributions in Table 5.2 are the same as the univariate frequency distributions of Civil Disobedience and Education presented for the same data in Section 2.1. Total row and total column frequencies each add up to 50, which is the total number of cases presented in the lower, right-hand cell of the frequency table. Sometimes this total is called the *grand total*.

I will always use the independent variable as the column variable and the dependent variable as the row variable, and I urge you to do the same this semester. Tables can certainly be set up the reverse way, but it is helpful if we always set them up the same way while learning cross-tabulation. Later on, when you become more familiar and comfortable with percentage tables, you can format them with independent variable values in the rows and dependent variable values in the columns. Meanwhile, do it my way.

5.2   Bivariate Percentage Tables

Frequency tables are OK (they do condense information in useful ways), but they are of limited direct use in determining if two variables are associated. Gauging an independent variable's effect on a dependent variable is difficult with frequency tables because values of the independent variable usually have different total numbers of cases. In the previous section, for example, we found that only 6 cases with less than a high school education believe in always obeying the law, compared with 11 high school graduates. But, of course, there are fewer cases with less than high school educations (10) than there are high school graduates (24), so it is not surprising that fewer cases with less education believe the law should always be obeyed. Who is more *likely* to believe the law should always be obeyed? It is hard to tell, or at least it would be if our table had hundreds or thousands of cases. What we need is some way to make comparisons that take into account the total number of cases in independent variable categories.

The way to handle this problem is probably obvious to you: percentaging. As we learned for univariate tables, percentaging is a way to standardize distributions regardless of how many cases there are. For a bivariate table, percentages equate distributions by giving each independent variable value the same total of 100 percent. Percentages tell us how many cases there would be in a cell *if* there were 100 cases with that value of the independent variable. By percentaging we can compare the Civil Disobedience distributions for respondents with less than a high school education, for high school graduates, and for respondents who have attended college, with each of these distributions standardized to a base of 100.

Thus, the fundamental rule for calculating percentages in a bivariate table: *Compute percentages within categories of the independent variable.* This rule is extremely important, so learn it well. This rule means that percentages will sum to 100 percent within each category of the independent variable. For tables set up with the independent variable in the heading across the top (as we are doing), the percentaging rule leads to calculating each percentage using the column total as the base of the percent. In other words, just divide each cell frequency by the total for that column, and then multiply by 100.

An example from the table in the previous section: For the 4 cases with less than a high school education who responded Follow One's Conscience, we get $\frac{4}{10}$ (100) = 40 percent. So, 40 percent of the least educated cases believe one should always follow one's conscience. For the 13 high school grads who have Follow One's Conscience scores, $\frac{13}{24}$ (100) = 54 percent, and so on.

Converted into a percentage table, the frequencies in Table 5.2 become the percentages in Table 5.3. We see from the table that, yes, civil disobedience certainly is related to education. Only 40 percent of less educated respondents believe that people should follow their consciences, compared with 54 percent of high school graduates and 75 percent of college-educated cases. Among these 50 cases, respondents with more education are more likely to believe that people should follow their consciences even if it means breaking the law and they are less likely to believe that people should always obey the law.

Note that each column in Table 5.3 is a univariate percentage distribution of the dependent variable for cases with that column's value of the independent variable. For example, the first data column is the univariate distribution of Civil Disobedience for persons with less than a high school education. As with all univariate distributions, percentages in each column will always add up to 100 (allowing, of course, for rounding error). This is a way to check the accuracy of your percentaging.

Table 5.3. Civil Disobedience by Education (in percentages)

Civil Disobedience	Education		
	Less Than Hi School	High School	College
Conscience	40	54	75
Obey Law	60	46	25
Total	100	100	100
(N)	(10)	(24)	(16)

Incidentally, I have not included a column of percentaged row marginals on the right-hand side. There is usually no need for such a column since cell percentages are based on column totals, not row totals.

There are other ways that tables can be percentaged, but they are not used very often in examining bivariate relationships. One could percentage within categories of the *dependent* variable so that percentages total 100 within each *row* category. Or one could use the *grand total* as the base for percentages so that the sum of *all* percentages is 100. Most standardized computer packages such as MicroCase permit these usually less useful ways to percentage, so you should be aware of them. They do have some specialized applications.[1] Usually, however, we want to compare the distributions of the dependent variable across different categories of the independent variable, and the only way to do this is to percentage within categories of the independent variable in order to eliminate the effects of different numbers of cases in those categories.

Table 5.3 is described as a 2 by 3 table because it has two rows and three columns, but in principle tables can have any number of rows or columns. Tables can be 2 by 2, or 3 by 2, or 3 by 3, or 3 by 4, and so on. In general, we refer to an r by c table, where r is the number of rows and c is the number of columns (not counting either the row or the column labeled Total).

But although tables theoretically can be of any size, it is very difficult to interpret large tables. Tables with more than, say, 12 to 16 cells are usually difficult to read. Therefore, you should keep tables as small as possible without losing too much information or obscuring

[1] For example, percentaging on the total number of cases is useful in so-called mobility tables showing the relationship between, say, the occupations of parents and children. Mobility tables are usually set up with parent's occupation in columns and child's occupation in rows. With all percentages based on the grand total, percentages in the diagonal can then be summed to find the percentage of respondents who are nonmobile. Percentages above and below the diagonal can be summed to find the percentages upwardly and downwardly mobile.

any important patterns in the relationship. Combine categories sensibly in constructing a table. The same guidelines for collapsing variables presented in Section 2.5 apply here as well. Combining adjacent values is particularly useful—and sometimes necessary—if some categories have very few cases. That is one reason I collapsed the Education variable even before beginning this example.

A reminder: Although the example used here has no missing data, any missing data should be excluded when you calculate percentages, just as missing data is excluded from univariate analyses.

5.3   How to Read Percentage Tables

I have already read Table 5.2, but now I want to describe in general terms how to interpret a table. We interpret a percentage table by making comparisons in the opposite direction from the way percentages are calculated. The general—and extremely important—rule for reading a table: *Compare percentages across categories of the independent variable.* This rule is important, chant it aloud while going between classes until you have memorized it . . . or, better yet, until you have *internalized* it.

We compare percentages by examining differences between percentages across independent variable categories. Large percentage point differences indicate a strong relationship between variables. Moderate percentage point differences mean a moderate relationship. And no percentage point differences imply no relationship between the variables. The latter situation is extreme and rare, however, and we usually take very small percentage point differences to indicate lack of a relationship.

But just how large is a "large" percentage point difference? What sort of difference is "moderate"? And how small is a "small" percentage point difference? These are tough questions since there are no hard and fast rules for assessing differences. As a rough rule of thumb, differences less than 10 percentage points are usually regarded as small, differences between 10 and 30 percentage points are moderate, and differences greater than about 30 percentage points are large. In Table 5.3 there is a 35 percentage point difference between college grads and those who did not graduate from high school (60 − 25 = 35). Yes, a strong relationship.[2]

---

[2] Please be careful to refer to *percentage point* differences, not *percentage* differences, when you describe the strength of relationships. What we are interested in is the difference between percentages, and that difference is measured in percentage points. There is a 35 *percentage point* difference between 25 and 60 percent. This is an important *point*. (The simple *percent* difference between 25 and 60 is 140—60 is 140 percent of 25.) Be careful to say exactly what you mean.

So, the smaller the differences between percentages across categories of the independent variable, the weaker the relationship. The larger such differences, the stronger the relationship.

We cannot, of course, rationally assess percentage point differences by mechanically applying the rough 10/30 percentage point guidelines. We need to *think* about the magnitudes of differences. We must take into account the marginal distributions of the dependent variables since badly skewed variables may limit possible differences among percentages. The substantive research problem we are studying and findings from previous research involving the same or comparable variables may provide a context for assessing differences among percentages.

Moreover, the theories and other ideas that initially shape our research may lead us to expect percentage point differences of a certain size. We can then gauge actual differences by comparing them with our expectations. However, in the last analysis (no pun intended) even reasonable men and women may disagree about the importance of a particular difference between percentages. I am *not* suggesting that assessment of percentage differences is arbitrary, only that there is room for some differences in interpretation.

Tables 5.4, 5.5, and 5.6, based on General Social Survey data, illustrate relationships of various strength involving gender as the independent variable. Table 5.4 reports very little difference between the percentages of males and females who believe in an afterlife—no relationship. In Table 5.5 we find a difference of 13 percentage points between males and females who report they own a gun—a moderate relationship. Table 5.6 shows a difference of nearly 30 percentage points between men and women who are afraid to walk in the area near where they live—a strong relationship.

Table 5.4. Belief in an Afterlife by Gender (in percentages)			Table 5.5. Gun Ownership by Gender (in percentages)			Table 5.6. Fear of Walking in Neighborhood by Gender (in percentages)		
After-life?	Gender		Own Gun?	Gender		Fear Walking?	Gender	
	Males	Females		Males	Females		Males	Females
Yes	82	83	Yes	48	35	Yes	26	55
No	18	17	No	52	65	No	74	45
Total	100	100	Total	100	100	Total	100	100
(N)	(759)	(978)	(N)	(848)	(1066)	(N)	(846)	(1057)
NO ASSOCIATION			MODERATE ASSOCIATION			STRONG ASSOCIATION		

By the way, finding no association is perfectly OK, so don't be disheartened when your analyses show no relationship between variables that you expected to be associated. So-called *negative findings* that contradict our expectations are as important as positive findings. They help us rule out possible causal connections. They tell us that some plausible ideas we had are not supported by the evidence, and science welcomes such falsifications. We might be understandably disappointed that data do not support our initial ideas, but we learn from negative findings. True, most published scientific studies report moderate or strong associations, but that is a publication bias that has little logical justification in the philosophy of science.

For 2 by 2 tables like 5.4, 5.5, and 5.6, the possible comparisons to make are limited and thus obvious within each table. One need compare only the percentages of males and females responding Yes (or, alternatively, the percentages responding No). Many percentage comparisons can be made, however, for tables larger than 2 by 2. Even a 2 by 3 table like Table 5.2 for Civil Disobedience by Education permits three comparisons: less educated with high school grads; less educated with college grads; and high school grads with college grads. But although you should inspect comparisons carefully, don't try to discuss every percentage comparison in large tables. Be selective and discuss only those that best describe or highlight the relationship. In other words, you should identify substantively significant and interesting patterns in the relationship between the independent and dependent variables, and then report those percentage point differences that best illustrate the patterns you have found. Your job in describing a table is to guide the reader through an interpretation of the data.

5.4   Positive, Negative, and Curvilinear Relationships

The directionality of ordinal and interval/ratio variables allows us to describe relationships as positive, negative, or curvilinear. A *positive relationship* is one in which higher scores on one variable are associated with higher scores on the other variable. Height and weight are positively associated, for example, since taller people tend to be heavier and shorter people tend to be lighter. We don't expect *perfect* patterns of this sort, of course. After all, there are some very heavy short people like Roseanne and some very skinny tall people like Abe Lincoln. What matters is the general pattern of the relationship.

Treating Follow Conscience scores as higher than Obey Law scores, we have already seen a positive relationship—that between Civil Disobedience and Education. Persons with more education tend to endorse civil disobedience more than do persons with less education. Table 5.7 presents another positive relationship—this one between education and respondents' descriptions of their personal financial situations as getting better, staying the same, or getting worse. These data are from the General Social Survey and make use of all five categories of educational level. The more education a respondent has, the better his/her perceived financial situation. Note how percentages increase steadily as we read across the first row and decrease steadily as we read across the bottom row. The fancy adjective that describes these kinds of ever-increasing or ever-decreasing patterns is *monotonic*. We don't demand such perfectly monotonic patterns as Table 5.7 displays, but positive relationships do show this general sort of pattern.

Table 5.7. Change in Financial Situation by Respondent's Education (in percentages)

Financial Situation	Education of Respondent				
	> HS	Hi School	Jr College	BA	Grad
Better	23	39	43	47	49
The Same	52	39	38	35	34
Worse	25	22	19	18	17
Total	100	100	100	100	100
(N)	(445)	(1562)	(186)	(470)	(223)

In a *negative relationship*, higher scores on one variable are associated with lower scores on the other variable. For example, amount of television watching is negatively associated with education. Table 5.8, based on the General Social Survey, presents the relationship between television watching (which I have collapsed into four categories) and education. Note that percentages decrease as we move across the top row and increase along the bottom row. Respondents with more education tend to watch television less; respondents with less education tend to watch television more. So, for example, 29 percent of respondents with less than a high school education watch television at least 5 hours a day. About 17 percent of high school graduates and only 11 percent of junior college graduates watch that much television. These percentages decrease to only 7 and 4 percent for respondents with undergraduate and graduate degrees. Similarly, percentages watching television 1 hour a day or less increase steadily from 14 percent for the least educated to 42 percent for the most educated. In short: The more education, the less TV watching. A clear negative relationship.

Table 5.8. Daily Television Watching by Education (in percentages)

Daily TV Watching	Education of Respondent				
	< HS	Hi School	Jr College	BA	Grad
5 or More Hours	29	17	11	7	4
3–4 Hours	35	34	28	25	20
2 Hours	22	29	37	30	34
0–1 Hour	14	20	24	38	42
Total	100	100	100	100	100
(N)	(293)	(1068)	(117)	(314)	(154)

As with positive relationships, we do not insist on such perfectly monotonic patterns before describing a relationship as negative, but we do require a general pattern of this sort. By the way, don't confuse negative *relationships* with the negative *findings* discussed in Section 5.3. The latter pertain to whether analyses support expectations about relationships, whereas negative relationships describe the direction of relationships.

Curvilinear relationships may take a variety of forms, but the simplest are relationships in which cases with low and high scores on the independent variable are similar in their scores on the dependent variable. Relationships that start positive turn negative, and relationships that start negative turn positive. Tables with curvilinear relationships have patterns shaped at least roughly like the letter V or an upside-down V.

Education and feelings of closeness to one's neighborhood have a curvilinear relationship, with both lower and higher levels of education associated with feelings of closeness to neighborhood. Table 5.9 describes this relationship. (I have dichotomized closeness to neighborhood to clarify the relationship.) Greatest closeness to neighborhood is reported by respondents who did not complete high school and respondents with college degrees. High school graduates and especially junior college respondents report less closeness. Yes, definitely a curvilinear relationship.

Table 5.9. Feelings of Closeness to Neighborhood by Education
         (in percentages)

Closeness to Neighborhood	Education of Respondent				
	< HS	Hi School	Jr College	BA	Grad
Feel Close	62	55	49	60	59
Not Close	38	45	51	40	41
Total	100	100	100	100	100
(N)	(183)	(740)	(81)	(220)	(101)

For ordinal or interval/ratio variables in all tables in this text, I have arranged values of the independent (i.e., column) variable from low on the left to high on the right, and values of the dependent (i.e., row) variable from high at the top to low at the bottom. If we follow this convention, in a positive relationship, cases tend to cluster along the *minor diagonal* of the table running from the lower left to the upper right (/). In a negative relationship, cases tend to cluster along the *major diagonal* from the upper left to the lower right (\). Curvilinear relationships in their simplest form are shaped like either a V or an upside-down V.

If variables' values are arrayed from left to right and from bottom to top, then positive, negative, and curvilinear patterns run like those in Figure 5.1, with Xs representing concentrations of cases reflected in relatively higher percentages within their columns. Since these are "ideal" patterns, we do not expect tables to conform to them exactly; we have to allow some latitude. And, too, we have to make allowances for tables that are not perfectly square since they cannot have perfectly straight-line diagonals or patterns quite V-shaped. Still, the patterns of positive, negative, and curvilinear relationships resemble the generalized shapes in Figure 5.1. In Chapter 10 we will encounter graphic analogs to these tabular patterns when we learn about scatterplots for interval/ratio variables.

Note that the patterns shown in Figure 5.1 also suggest graphically why we cannot speak of positive or negative relationships if one or both variables are nominal. Imagine what would happen to the patterns of Xs if we shuffled categories of the independent or dependent variables, as we certainly can do for nominal variables. Descriptions of variable values as ranging from low to high would be meaningless, and the straight lines of Xs would become scrambled. Directionality makes no sense for nominal variables because the ordering of their values is arbitrary. We cannot speak, for example, of the values of religion or sex or race as "lower" or "higher."

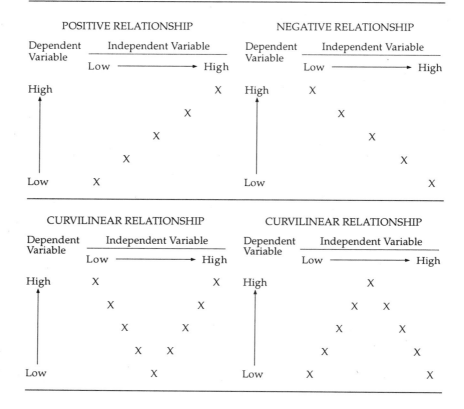

Figure 5.1. Patterns of Positive, Negative, and Curvilinear Relationships

5.5   Format Conventions for Bivariate Tables

As with univariate tables, there is no single, universally agreed upon format for bivariate tables. Disciplines vary in format preferences, and there is further variation even within disciplines. Tables in this chapter conform to format conventions of the American Sociological Association. You can use them as models for your own tables.

Here are suggestions for preparing readable, presentation-quality bivariate tables consistent with the practices of most social scientists:

- Number your tables with arabic numerals.

- Use a clear, straightforward title that describes the contents of the table in this form: Dependent Variable by Independent Variable (e.g., Civil Disobedience by Education). The title also may identify the data source or the set of cases to which the table pertains if this information is not clear in your narrative accompanying the table.

- Label the leftmost column with the name or description of the dependent variable (e.g., Civil Disobedience) and the other columns with the values of the independent variable (e.g., values of Education). Above the independent variable's values, enter the name of the independent variable, with a horizontal line below extending over the variable's values.

- Include a Total row that adds column percentages as a guide to anyone reading the table. Sometimes, percentages do not total exactly 100 percent because of rounding errors. In such cases either enter the exact total (e.g., 99.9 or 100.1) or enter 100.0 percent with a footnote below the table indicating that percentages may not sum to exactly 100 percent due to rounding errors.

- Include an (N) row presenting the number of cases on which column percentages are based. This N allows the table's reader to calculate cell frequencies and then combine categories in a different way if he/she wishes. Report these totals in parentheses at the bottom of the independent variable columns. It is *not* conventional to report marginal totals for the dependent variable, so you should leave them off your table unless there is some good reason to present them.

- Retain only significant digits in percentages. Usually you should round percentages to either whole numbers or one decimal place. This guideline has exceptions, but not many when working with social scientific data.

- Use the same number of decimal places in all percentages.

- Do not put individual cell frequencies in the table—just percentages. As with univariate percentages, an interested table reader can recover any frequency by multiplying the column N by the percentage and then dividing by 100.

- Keep columns of percentages equal distances apart and right-adjust column percentages. (Better yet, align decimal points if your word processor has that feature.)

- Do not put % signs after cell entries. As with univariate tables, % signs in bivariate tables are unnecessary, clutter a table, and are tacky, so leave them out.

- Do not draw vertical lines in a table. They too clutter. Guide the reader's eye and provide definition for the table simply by drawing a horizontal double line between title and column headings and horizontal single lines below column headings and at the bottom of the table. Examples of these lines are shown in the tables throughout this textbook.

- As always when you present statistical analyses to others, be very neat. Keep cell entries lined up, horizontal lines the same length, and so on. As I suggested earlier, you owe both style and substance to readers of your tables.

Each table in this chapter illustrates a proper way of formatting a bivariate table. Notice that for an ordinal or interval/ratio independent variable (like the education variable), values of the independent variable usually are listed from lowest at the left to highest at the right. For the dependent variable (like TV watching), it is helpful to list values from the highest at the top to the lowest at the bottom. This ordering allows ready recognition of positive and negative patterns described in the previous section and, we will see in Chapter 9, is analogous to the way that graphic displays of interval/ratio variables are set up. However, this arrangement of values is not a strong convention and there are many exceptions to it, especially with regard to the ordering of dependent variable (row) values. Moreover, MicroCase and other statistical computer programs do not always order variables (especially interval/ratio variables) this way. A program's ordering of values depends on how the variables' values are coded in the data file. In fact, programs typically order values of the row variable from the lowest at the top to the highest at the bottom. Therefore, you may want to adjust the ordering of values and accompanying percentages when you transform computer output into presentation-quality tables. These ordering conventions have no meaning for nominal variables, of course, since their values have no rank-order. You should list values of nominal variables in some reasonable, conventional way. Often it is clearest to list nominal categories from the most frequently to the least frequently occurring.

I have pointed out that in this text I am setting up bivariate tables with the independent variable in columns and the dependent variable in rows. Consistency in format reduces confusion. However, this convention is far from universal, and you should be able to read tables set up the other way around. For example, Tables 5.10 and 5.11 describing the relationship between opinion on spending to improve the condition of blacks and political party affiliation are identical except for their positioning of the independent and dependent variables. (In describing table format, I don't want to overlook the substance of these tables: Based on the General Social Survey data, Tables 5.10 and 5.11 show that Democrats are most sympathetic and Republicans are least sympathetic to spending to help African Americans.)

Table 5.10. Spending to Help Blacks by Political Party (in percentages)

Spending to Help Blacks	Political Party		
	Dem	Indep	Rep
Too Little	48.2	38.5	18.7
About Right	38.6	34.4	52.3
Too Much	13.2	27.1	29.0
Total	100.0	100.0	100.0
(N)	(581)	(192)	(493)

Table 5.11. Spending to Help Blacks by Political Party (in percentages)

Political Party	Spending to Help Blacks				
	Too Little	About Right	Too Much	Total	(N)
Dem	48.2	38.6	13.2	100.0	(581)
Indep	38.5	34.4	27.1	100.0	(192)
Repub	18.7	52.3	29.0	100.0	(493)

5.6 Stacked Bar Graphs for Bivariate Relationships

A variant of bar graphs called *stacked bar graphs* offers an especially efficient way to visually describe bivariate relationships. Figure 5.2 is a stacked bar display of the relationship between opinions on spending to help African Americans and political party. We just saw this relationship in Tables 5.10 and 5.11, but the stacked bar display more quickly conveys information about this relationship. Each bar is divided to show the percentage of that particular political group that believes spending on blacks is too little, about right, or too much. Depending on whether the stacked bar graph reports actual percentages (MicroCase does not), the stacked bar graph's information may lack some of the detail found in tables, but we get a much faster and more vivid overall impression of the relationship between variables. The graph in Figure 5.2 immediately draws our eye to the declining "steps" of the lower bar segments, telling us that Democrats support spending more than Independents, who favor spending to help blacks more than do Republicans. We also readily see that Independents and Republicans don't differ much in percentages who believe too much is spent on helping blacks.

No doubt about it: style matters. That's why you dress up for job interviews, that's why you sport a cool doo, and that's why you need to consider using stacked bar graphs if they better convey the information you want to analyze or pass on to others. Be careful, though. Sometimes graphs dazzle without informing. (Turn on TV news or visit a newsstand for all the examples you want.) You'll need to carefully decide whether tables or graphs (or sometimes both) best express the information you need to consider or convey.

Figure 5.2.   Opinion About Spending to Help Blacks by Political Party

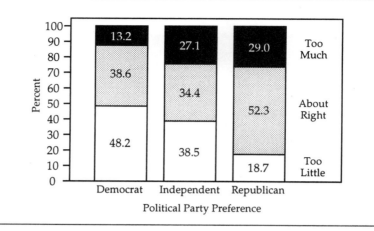

5.7   A Caution About Bivariate Tables Based on Small Ns

Be careful with bivariate tables in which one or more column N is small. Generally speaking, the larger the total frequencies in the independent variable's categories, the more stable and reliable the percentages and thus the more confidence we have in them. That is, the shift of a few cases from one cell to another would not affect percentages very much. If percentages are based on a very small column total, however, the shift of just a few scores from one dependent variable value to another can radically change the percentage distribution. The individual column (i.e., independent variable) totals matter much more than the grand total since they are what cell percentages are based on.

We already learned this importance of an adequate N for univariate distributions in Section 2.3. It is no surprise that the same principle applies to columns of bivariate tables since each of these columns is a univariate distribution for a given value of the independent variable. As a very loose—and, admittedly, often violated—rule-of-thumb, percentages should be based on at least, say, 30 cases. Indeed, some statisticians urge at least 50 or even 100 cases. (Yes, I know that the 50-case, Education → Civil Disobedience data in this chapter come close to violating this guideline. We stat teachers are authorized to break a few rules now and then.)

If categories of an independent variable have Ns that are too small for stable percentages, consider collapsing or deleting values in some reasonable fashion. Interval/ratio variables like age in years and income in dollars often have many values and thus are particularly good candidates for collapsing in a cross-tabulation. If the variable is nominal, you may also be able to handle small column totals by excluding the problem value(s). For example, type of residence, a variable in the full General Social Survey, has only 14 respondents (out of a total N of 2210) who report that they live in commercial housing like hotels. In contrast, 218 GSS respondents live in trailers and 1700 in single-family homes. In an analysis using residence type as a variable, it may make sense either to entirely exclude these 14 cases who live in commercial housing or to combine them with the Other category (which would helpfully boost Other's small N from 30 to 44).

Although only the totals of independent variable categories affect the stability of percentages, at times you may also find it useful to collapse (sensibly, of course) dependent variable values. If many uncollapsed cells have small percentages that tell little about the relationship between the variables, collapsing values may help us identify and interpret relationships. We will also find in the next chapter that some rigorous analysis procedures require that we expect at least a certain number of cases in each cell. Collapsing categories may help meet that condition.

5.8   Association Does *Not* Imply Causation

This brief section is much more important than it may look, so read it very carefully.

A relationship between variables does not necessarily mean that one variable causes the other. Two variables may be associated with one another even if they are not *causally* related. Many associations that are not causal are due to concurrent but independent trends or changes over time. The universe is expanding and you are getting older, but your getting older does not cause the universe's expansion . . . nor does the universe's expansion cause your getting older. Your age and the universe's size vary together independently of one another. Likewise, you are accumulating college credits and the world's population is growing, but neither is causing the other.

Many noncausal associations are due to the relationships of variables to other variables besides the passage of time per se. Here are some classic examples frequently cited in statistics or research methods books:

- The number of storks is associated with birth rates in regions of Europe. (Regions with more storks have higher birth rates.)

- The flooding of the Ganges River in India is associated with the rate of street crime in New York City. (When the Ganges River overflows its banks, street crime increases in New York City.)

- The average salary of clergy is associated with liquor consumption. (Cities and towns that pay priests, ministers, and rabbis more have higher liquor consumption.)

- The number of firefighters at a fire is associated with the fire's property damage. (The more firefighters at a fire, the greater the property loss.)

- Shoe sizes of children are related to their scores on spelling tests. (The bigger kids' feet, the better they spell.)

So storks bring babies? Flooding in India causes muggings in New York City? Well-paid clergy spend their money on booze? Firefighters cause property damage? Big feet improve test performance? All these conclusions are pretty doubtful. In fact, they are ridiculous. What is happening is that additional variables are causing each of these relationships, making it seem as if the two variables are causally connected even though they are not.

With storks and babies, for example, it seems likely that rural areas have more storks than urban areas, and rural areas also have higher birth rates than urban areas (for reasons quite unrelated to the presence or absence of storks). Ruralness thus causes both lots of storks and high birth rates, creating a statistical association between these two variables that is not genuinely causal. The variables are statistically associated, but the relationship is not causal. I'll let you figure out what is "really" happening with the Ganges flooding–street crime, the clergy salaries–liquor consumption, the firefighters–property loss, and the foot size–spelling relationships.

Association is, of course, crucially relevant to causation. Causally related variables are associated. But the converse isn't true. Variables that are associated are not necessarily causally related. Like all rational people, social scientists avoid falsely attributing causality to an association by insisting on theoretical rationales for linking variables. Here is where theory—*good* theory—comes in. There must be some good reason why one variable causes another before we describe them as causally related. Emile Durkheim, for example, didn't merely associate religion and suicide—he explained *why* they are linked. In the absence of a good reason why one variable causes another, an association may be interesting, even provocative, and we certainly should study why variables are associated in order to uncover one or more reasons for their linkage. But until a reason is found, let's not describe the association as causal.

In his *Mismeasure of Man*, Stephen Jay Gould suggests that equating association and causation "is probably among the two or three most serious and common errors of human reasoning." Let's avoid this fallacy. Let's be extremely cautious not to commit the fallacy of assuming that two variables that are associated are necessarily causally related. Maybe they are . . . and maybe they aren't. We will take these alleged relationships up again in Chapter 11, where we will learn that such statistical associations that are not really causal relationships are called *spurious relationships*. Meanwhile, remember: Association does *not* necessarily imply causation.

5.9   Summing Up Chapter 5

Here is what we have learned in this chapter:

- Bivariate cross-tabulation is one of several techniques for determining if two variables are associated.
- Percentages should be calculated *within* categories of the independent variable.
- Percentages should be compared *across* categories of the independent variable.
- Missing data values should usually be excluded from a bivariate percentage table.
- Bivariate percentage tables for ordinal or interval/ratio variables reveal the direction of a relationship—positive, negative, or curvilinear.
- Bivariate percentage tables should be presented using conventional formats that facilitate reading the table.
- Stacked bar graphs visually display bivariate relationships.
- Percentages should not be based on small Ns—Ns of, say, less than 30.
- A statistical association between two variables does not mean they are *causally* related. Association does *not* necessarily imply causation.

Before we continue learning about bivariate relationships, let's briefly take stock of where we are in learning statistics. In the first four chapters, we learned ways to describe individual variables—univariate analysis. Now we have begun to learn ways to describe the relationship between two variables—bivariate analysis. In particular, we have learned how to answer the first three questions posed in the introduction to this chapter:

a.   Is there a relationship between the two variables in the data we are analyzing?

How to Answer: Compare percentages across categories of the independent variable. Differences in percentages indicate that the variables are associated with one another.

b. How strong is the relationship?

How to Answer: Assess the magnitude of the differences in percentages across independent variable categories. The larger the differences, the stronger the relationship.

c. What is the direction and shape of the relationship?

How to Answer: Examine the pattern of percentages to identify positive, negative, or curvilinear relationships.

As we learn more statistics, we will find additional ways to answer these three questions about relationships between variables. Nevertheless, cross-tabulation will remain an important tool in our statistical toolbox. In the next chapter we'll find how to generalize bivariate relationships from sample data to a larger population.

By the way, results of statistical methods described in the next two chapters are usually integrated with discussions of bivariate percentage tables, so I will wait until we finish Chapter 7 before offering examples of analysis write-ups for cross-tabulations.

Key Concepts and Procedures

Ideas and Terms

cross-tabulation (cross-tab or cross-classification or tabular analysis)
research hypothesis
independent variable
dependent variable
bivariate frequency distribution
joint frequency (cell frequency)
marginal distribution

grand total
negative findings
monotonic
positive relationship
negative relationship
curvilinear relationship
major and minor diagonals
stacked bar graph

Symbols

r
c
r by c

CHAPTER 6
The Chi-Square Test of Statistical Significance

In this chapter we take up tests of statistical significance—ways of deciding whether a bivariate relationship found in sample data is also likely to be found in the population from which the sample was drawn. Our special interest will be in the chi-square test of statistical significance applied to bivariate cross-tabulations. At times this chapter necessarily involves thinking abstractly, so you may have to read it carefully a couple times.

After this chapter you will be able to:

1. Explain the logic of tests of statistical significance.

2. Explain and offer examples of null hypotheses.

3. Explain the difference between Type I and Type II errors and offer examples of each.

4. Carry out and interpret chi-square tests on cross-tabulations.

5. Use a chi-square table and interpret chi-square values in terms of statistical significance.

6. Recognize difficulties using chi square with expected frequencies less than 5 and handle such difficulties by excluding or collapsing values.

7. Distinguish between statistical significance and substantive significance; explain the relationship of statistical significance to sample size and the strength of a relationship; and exercise caution in interpreting results of chi-square tests when Ns are large.

8. Recognize, again, that association does not imply causation.

6.1   The Logic of Tests of Statistical Significance

Tabular analysis procedures discussed in Chapter 5 answer the question: Is there a relationship between two variables in the data we are analyzing? But if we find a bivariate relationship and if the data are from a sample rather than an entire population, we need to ask a second question: Is there a relationship in the larger population from which the sample was selected?

After all, it might happen just by chance that the particular cases selected show a relationship between two variables even if there really is no relationship in the population as a whole. Maybe, for example, there really is no relationship between education and attitude toward civil disobedience among all adult Americans, but the General Social Survey or the 50-case subsample we are using just happened to select respondents among whom the less educated favor always obeying the law and the more educated favor following one's conscience. As we saw in the IQ example in Section 4.5, had the GSS randomly chosen different respondents, maybe we would not have found a relationship between the variables Education and Civil Disobedience. After all, samples vary.

(Here the reasoning becomes a little tricky, so concentrate very hard on the rest of this section, especially the next two paragraphs. Read them more than once.)

What we need is some way to determine the probability (i.e., the odds) that we would find a relationship in our sample if there were no relationship in the population. If that probability is small (say, less than 1-in-20), we can conclude that there probably is a relationship in the larger population. After all, if there is only a small chance of finding a relationship in the sample if there is no population relationship, and yet we find a sample relationship, then there is probably a relationship in the population too. Statisticians use the term *null hypothesis* (often symbolized H_0) to refer to the assumption of no relationship in the population. In statistical jargon: We *reject the null hypothesis* of no population relationship. There probably *is* a relationship in the population, and we say the relationship is *statistically significant*.

On the other hand, if the odds are pretty high (say, more than 1-in-20) that we would find a relationship in the sample even if there were no relationship in the population, then we cannot have much confidence that there really is a relationship in the larger population. The relationship we find in the sample might have occurred just by chance. In this case, we *fail to reject the null hypothesis* of no relationship in the population. There might well be no relationship in the population, and we say the relationship is *not* statistically significant.

The probability that a relationship found in sample data occurs by chance even if there really is no relationship in the population sampled is called the *level of statistical significance*, or just *significance level*. Significance level is expressed as a number ranging from 0 to 1.00. That is the conventional way of expressing probabilities. A *probability* indicates the odds or chance of something happening, in this case the odds of finding a sample relationship even if there is no population relationship.

More specifically, a probability multiplied by 100 (to clear decimal places) is the number of times an event is likely to occur out of 100 trials. A probability of 0 means that the chance of something happening is nonexistent—it is impossible; it will never occur. Hell freezing over has a probability of .00. Hell won't freeze over in 100 winters. Something that occurs 1 out of 100 times has a probability of .01. Something that occurs, say, 1 out of 20 times has a probability of .05. A probability of .50 means that something is likely to occur about 50 out of 100 times—like getting a head on a flip of a coin. A probability of .95 describes something highly likely to occur, but not for sure. A probability of 1.00 means that something is completely certain—a sure bet; it is likely to occur 100 out of 100 times. (There are not many of those in life, are there?) So, the smaller the probability of something, the less likely it is. The higher the probability, the more likely.

I have suggested that statistical significance depends on whether the odds of finding an observed relationship are 1-in-20 or less. There is a reason I chose 1-in-20 as a cutoff point for deciding on statistical significance. By convention, researchers regard a probability of .05 (i.e., 1-in-20) or smaller as reasonable odds for determining statistical significance. They report such statistical significance as $p < .05$ (with p standing for probability and < meaning less than). Naturally, researchers report even lower levels when they are found since those odds are even smaller. (Levels of significance with such low probabilities are commonly rounded up to .01 or .001, with probabilities reported as $p < .01$ or $p < .001$.) But the .05 level is widely used as a cutoff for statistical significance. If the probability of getting an observed relationship in sample data is .05 or less, and yet we observe the relationship, we feel reasonably confident that the relationship "really" exists and is not due merely to chance. After all, .05 or less is a pretty small probability of finding a relationship in a sample if there is no relationship in the population, so we can reject the null hypothesis of no relationship.

Keep in mind, though, that there is nothing absolute or fixed about this .05 level. It is just the level conventionally used to decide on statistical significance. Not all researchers and statisticians are pleased with this convention. In fact, some have described it sarcastically as the "sacred" .05 level. But the .05 level is a cultural creation, of course, and is not handed down by some statistically minded deity. (Judging from the Old Testament, God is probably a lot more demanding.) Actually, the .05 level was handed down (so to speak) by Sir Ronald Fisher in his influential 1925 book *Statistical Methods for Research Workers*. Many disciplines and subdisciplines employ cutoff levels of significance other than .05. Physicists who study subatomic particles, for example, apply the far more stringent probability of .0001 or less before they acknowledge detection of a new particle. But we social scientists usually use the less stringent .05 level.

The fact that statistical significance is based on probability means that we can never be absolutely certain we are right when we either reject or fail to reject a null hypothesis. After all, we never know for sure from sample data whether or not two variables are related in the population. Only population data can tell us that with absolute certainty. Errors are always possible with sample data because our sample may be unrepresentative. Even random sampling can result in an unrepresentative sample (although it usually doesn't).

When we reject a null hypothesis that is really true, we commit what statisticians call a *Type I error* or an *alpha error*. The significance level is the probability that we are committing a Type I error if we reject the null hypothesis. The significance level is sometimes referred to as the *alpha level* and is symbolized as α. At the .05 alpha level, we will be wrong less than 1 time out of 20.

But we also commit an error if we fail to reject a null hypothesis that is really false. Demonstrating their ability to count even in roman numerals, statisticians call this a *Type II error* (or *beta error*). The probability of committing a Type II error using a significance test like the chi-square test is called the *power* of the test. The more powerful a significance test, the more likely we are to reject a false null hypothesis. Table 6.1 summarizes Type I and Type II errors (with H_0 standing for null hypothesis).

Table 6.1. Type I and Type II Errors

Decision About H_0	If H_0 Is True	If H_0 Is False
Reject H_0	Type I Error	No Error
Do Not Reject H_0	No Error	Type II Error

Since the probability of a Type I error is α, it is tempting to conclude that the probability of a Type II error is $1 - \alpha$. Wrong. An α of .05 does *not* mean that the probability of a beta error is .95. Beta errors are a little more complicated than that, and are beyond what we can discuss in this text. It is true, though, that Type I and Type II errors are inversely related and that reducing the chance of a Type I error increases the chance of a Type II error, and vice versa.

There is no way we can completely eliminate the possibility of committing an error when we decide either to reject or not to reject a null hypothesis. But we can determine the probability that we are committing a Type I error if we reject the null hypothesis since that is what the level of statistical significance tells us.

This is where the idea of sampling distributions introduced in Section 4.5 becomes especially useful. Sampling distributions allow us to find the significance level—that is, the probability that we would have found a relationship in a sample if the variables are unrelated in the population. Thus, sampling distributions allow us to decide rationally whether a relationship is statistically significant or not, and tell us the probability that we are committing an error if we reject the null hypothesis of no relationship in the population. Which sampling distribution and significance testing procedures are most appropriate depends on the characteristics of the data we are analyzing. For cross-tabulations, statisticians have devised a widely used test of significance based on the sampling distribution of a statistic called chi square. Let's see how to use this test.

6.2   The Chi-Square Test

Chi square, symbolized with the Greek χ^2 and pronounced $k\bar{i}$ square, is a number that compares the actual frequencies in a bivariate table with the frequencies we would expect if there were no relationship between the two variables in the larger population. The distribution of chi square is a very useful sampling distribution. If we assume that two variables are unrelated in a population, we can determine the probability of getting a chi square of a given size.

Here is the logic of the *chi-square test*: If *observed frequencies* in a bivariate table based on sample data are similar to the frequencies we would expect if there were no relationship between the two variables in the population, then we cannot reject the H_0 of no relationship. We can feel pretty confident that there is really no relationship in the population (thus risking a Type II error). On the other hand, if the observed frequencies in a table are quite different from those expected if the variables are unrelated, then we reject the assumption that the variables are unrelated—and so we conclude that there probably is a relationship between the variables (and risk a Type I error).

But how do we find *expected frequencies*? Let's take an example. Consider again the 50-case Education → Civil Disobedience relationship introduced in Section 5.1 and percentaged later in that chapter. Table 6.2 presents the bivariate frequency distribution for this relationship.

Table 6.2. Civil Disobedience by Education
            (in frequencies)

Civil Disobedience	Education			Total
	Less Than Hi School	High School	College	
Conscience	4	13	12	29
Obey Law	6	11	4	21
Total	10	24	16	50

If Education and Civil Disobedience were unrelated, then each education category would have the same proportion of cases who have Follow Conscience scores. Each Education category would also have the same proportion who have Obey Law scores. That is what it means to say that variables are unrelated: The proportion of less-educated people who responded Follow Conscience is the same as the proportion of high school graduates who responded Follow Conscience, and both are the same as the proportion of more-educated people who responded Follow Conscience. If these proportions are the same, the percentages will be the same, and a comparison of percentages across categories of the independent variable Education will find no differences. In fact, each of the Civil Disobedience distributions within Education categories would be proportional to the total distribution of Civil Disobedience, which is the marginal distribution in the right-hand column of Table 6.2. (This paragraph is important enough to warrant rereading.)

It is easy to find these expected cell frequencies—that is, the cell frequencies we would find if the variables were unrelated. Let's first consider the frequencies we expect in the cells of the first row. Since $\frac{29}{50}$ of all cases responded Follow Conscience, we expect $\frac{29}{50}$ of the 10 respondents with less than a high school education to have Follow Conscience scores. That's $\left(\frac{29}{50}\right)10 = 5.80$ respondents. Similarly, since $\frac{29}{50}$ of all respondents responded Follow Conscience, we expect $\frac{29}{50}$ of the 24 respondents with less than a high school education to have Follow Conscience scores. That's $\left(\frac{29}{50}\right)24 = 13.92$ respondents. And likewise for college-educated respondents: $\left(\dfrac{29}{50}\right)16 = 9.28$ respondents.

So, to find the expected frequency for each cell, all we do is this:

1. Divide each row marginal (i.e., row total) by the grand total N.
2. Multiply that number by the column marginal (i.e., column total).

As a formula:

$$f_e = \left(\frac{\text{Row Marginal}}{N}\right)\text{Column Marginal}$$

where f_e = expected cell frequency

N = total number of cases

That's how we find the expected frequency for each cell assuming that the variables are unrelated.

Applying the formula to find the expected number of less-educated cases who would have Follow Conscience scores if the variables were unrelated:

$$f_e = \left(\frac{\text{Row Marginal}}{N}\right)\text{Column Marginal}$$

$$= \left(\frac{29}{50}\right)10$$

$$= (.58)(10)$$

$$= 5.80$$

So there would be 5.80 cases (rather than the 4 observed cases) in the upper left-hand cell if Education and Civil Disobedience were unrelated.

To find the expected number of high school graduates who responded Follow Conscience:

$$f_e = \left(\frac{29}{50}\right)24$$

$$= (.58)(24)$$

$$= 13.92$$

If the variables were unrelated, we would find 13.92 cases (rather than the 13 observed cases) in the High School–Follow Conscience cell.

I'll leave it to you to find the expected number of college-educated respondents who believe that people should follow their consciences. (I'm counting on you to come up with an expected frequency of 9.28.)

Table 6.3 is the frequency table for this relationship with expected cell frequencies in parentheses. Add the expected frequencies and you will find that they total both the row and column marginals. For example, 5.80 + 13.92 + 9.28 = 29, and 5.80 + 4.20 = 10. In other words, the expected frequencies preserve the original marginal distributions:

Table 6.3. Civil Disobedience by Education
(in frequencies and expected frequencies)

Civil Disobedience	Education			Total
	Less Than Hi School	High School	College	
Conscience	4 (5.80)	13 (13.92)	12 (9.28)	29
Obey Law	6 (4.20)	11 (10.08)	4 (6.72)	21
Total	10	24	16	50

Furthermore (and unsurprisingly if you have been following the logic of what we are doing), we find the variables unrelated if we percentage the expected frequencies. That is how it should be since the expected frequencies assume that the variables are unrelated. These percentages based on expected frequencies are shown in Table 6.4. We never present this kind of table in reports of our statistical analyses, but I am showing it to you so you can see clearly that expected frequencies are the frequencies we would find if variables were unrelated.

Table 6.4. Civil Disobedience by Education Based on
Expected Frequencies (in percentages)

Civil Disobedience	Education			Total
	Less Than Hi School	High School	College	
Conscience	58	58	58	58
Obey Law	42	42	42	42
Total	100	100	100	100
(N)	(10)	(24)	(16)	(50)

Now we are ready to compute chi square. We use the following formula:

$$\chi^2 = \Sigma \frac{(f_o - f_e)^2}{f_e}$$

where χ^2 = chi square

f_o = observed frequency in each cell

f_e = expected frequency in each cell

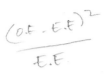

We have encountered Σ (sigma) many times already. Once again, it tells us to add everything that comes after it.

So to find chi square, first calculate the expected frequency for each cell and then follow the formula:

1. Subtract the expected frequency from the observed frequency for each cell.

2. Square each difference.

3. Divide each squared difference by the expected frequency for that cell.

4. Add all these numbers for all cells.

The result is χ^2. Don't forget to square each cell difference (step 2) *before* dividing by the expected frequency for that cell.

You can see from the numerator (i.e., top part) of the formula that the larger the difference between the observed and expected frequencies, the larger χ^2 will be. Since expected frequencies assume no relationship, larger differences between observed and expected frequencies indicate a stronger relationship (i.e., a relationship further from no relationship). Thus the stronger a relationship (for a given number of cases), the larger the value of chi square.

Applying the chi-square formula to the observed and expected frequencies in the Education → Civil Disobedience example:

$$\chi^2 = \Sigma \frac{(f_o - f_e)^2}{f_e}$$

$$= \frac{(4 - 5.80)^2}{5.80} + \frac{(6 - 4.20)^2}{4.20} + \frac{(13 - 13.92)^2}{13.92} + \frac{(11 - 10.08)^2}{10.08} + \frac{(12 - 9.28)^2}{9.28} +$$

$$\frac{(4 - 6.72)^2}{6.72}$$

$$= \frac{3.2400}{5.80} + \frac{3.2400}{4.20} + \frac{.8464}{13.92} + \frac{.8464}{10.08} + \frac{7.3984}{9.28} + \frac{7.3984}{6.72}$$

$$= .5586 + .7714 + .0608 + .0840 + .7972 + 1.1010$$

$$= 3.373$$

You may find the mechanics of computing χ^2 even easier with a simple computational table like Table 6.5.

Table 6.5. Computation of Chi Square for Civil Disobedience by Education

f_o	f_e	$f_o - f_e$	$(f_o - f_e)^2$	$(f_o - f_e)^2 / f_e$
4	5.80	−1.80	3.2400	.5586
6	4.20	1.80	3.2400	.7714
13	13.92	−.92	.8464	.0608
11	10.08	.92	.8464	.0840
12	9.28	2.72	7.3984	.7972
4	6.72	−2.72	7.3984	1.1010
Sum				$\chi^2 = 3.3730$

At this point, statisticians have done something wonderful for us: They have figured out the sampling distribution of chi square. That is, they have calculated the probability of getting a chi square of at least a given size even if there is no relationship between two variables in the larger population from which the sample was drawn. This probability requires that the data we are analyzing are from a *random* sample of the population—an assumption met quite well by carefully drawn survey data like those of the General Social Survey.

The probability of getting a particular chi square depends in part on **degrees of freedom (df)**. Recall that we encountered the idea of degrees of freedom before when we learned about the variance and standard deviation for sample data in Section 4.1. Degrees of freedom for the chi-square test are equal to $(r - 1)(c - 1)$, where r and c are the numbers of rows and columns in the table (not counting Totals). So, degrees of freedom are 1 for a 2 by 2 table, 2 for a 2 by 3 table, 3 for a 2 by 4 table, and so on. In the above 2 by 3 table:

$$\begin{aligned} df &= (r-1)(c-1) \\ &= (2-1)(3-1) \\ &= (1)(2) \\ &= 2 \end{aligned}$$

The distribution of chi square is presented in Table 1 in the Appendix. The table shows the minimum value of chi square needed to achieve significance at the .05, .02, .01, and .001 levels. To use the table, find the degrees of freedom in the leftmost column and then read across to see how large a chi square is needed to reject the null hypothesis at a given level of significance. Here is the row of the chi-square table for 2 degrees of freedom:

	Probability			
df	.05	.02	.01	.001
.	.	.	.	.
2	5.991	7.824	9.210	13.815
.	.	.	.	.
.	.	.	.	.
.				

We see that with 2 degrees of freedom we need a chi square of 5.991 or more to achieve significance at the .05 level, a chi square of at least 7.824 for significance at the .02 level, 9.210 for the .01 level, and 13.815 for the .001 level.

Our chi square is only 3.373, less than the chi square associated with even the .05 level. Thus we fail to reject the null hypothesis of no relationship between Education and Civil Disobedience in the population from which these 50 cases were drawn. We could have gotten the frequencies we observe in our sample data just by chance even if there is no relationship between the two variables in the larger population. Since we could have gotten our sample relationship by chance, we don't feel confident rejecting the null hypothesis of no relationship. In statistical jargon, we "fail to reject the null hypothesis." Note that we never "accept" a null hypothesis. Rather, statistical reasoning proceeds by falsification—that is, not by accepting ideas, but by rejecting alternative ideas. This is the way science works (and, not so incidentally, it works extremely well).

Of course, we might be committing an error—a Type II error—in not rejecting the null hypothesis if it really is false. Still, the chances are more than 1 out of 20 that our observed relationship could occur by chance even if there really is no relationship in the larger population.

As we use statistics like chi square, it is easy to take them for granted and forget their special meanings. But, like our friends, statistics should not be taken for granted. I hope you will keep in mind that chi square is a statistic in a technical sense. That is, it is a characteristic of a sample, just like a mean or a standard deviation or a percentage can be characteristics of a sample. Also be mindful that the chi-square table in the Appendix describes a sampling distribution. It describes the distribution of a statistic (chi square) by reporting the probabilities associated with every possible sample outcome (i.e., every possible chi square). For the two variables we are analyzing, imagine every possible bivariate table with a given N, each with an associated chi square. The chi-square table in the Appendix shows the distribution of those chi squares by reporting a probability of getting each one if there were no population relationship. We then find where our particular relationship's chi square is in that distribution. That is, we find out how likely our chi square is to occur if there really is no relationship in the population. That probability is the level of statistical significance of our sample relationship.

Incidentally, researchers often report chi square and the associated probability immediately below a table, as in Table 6.6. Here that probability is greater than .05, indicating that the relationship is not statistically significant. The abbreviation n.s. stands for not significant. When relationships are statistically significant, probabilities are usually reported as $p < .05$, $p < .02$, $p < .01$, or $p < .001$, depending on the significance level of χ^2. Often, too, a measure of association of the kind we will learn in the next chapter is reported along with the chi-square test, but more on that later.

Table 6.6. Civil Disobedience by Education
(in percentages)

Civil Disobedience	Education		
	Less Than Hi School	High School	College
Conscience	40	54	75
Obey Law	60	46	25
Total	100	100	100
(N)	(10)	(24)	(16)

$\chi^2 = 2.00$; n.s.

Tests of statistical significance like the chi-square test are extremely helpful tools for analyzing data. They tell us how confident we can be that findings are "real" rather than just the result of chance or randomness in the sampling process. But all tools must be used with caution. (Ever hit your thumb with a hammer?) The next two sections describe major cautions about the chi-square test that we must keep in mind.

6.3   Problems with Expected Frequencies Less Than 5

I deliberately kept numbers small in the preceding example so that we could focus on the logic and procedures of a chi-square test. But let me admit a serious problem with these small numbers: We should be very cautious carrying out a chi-square test if any expected frequency is less than 5.0, and especially if more than about 20 percent of cells have expected frequencies that small. (The Education → Civil Disobedience example skirts quite close to violating this criterion, with an expected frequency less than 5.0 in one, or 16.7 percent, of the six cells.) If we violate the assumption of expected frequencies of at least 5.0, we are on weak grounds using the probability associated with chi square. The reason is that the distribution of chi square is assumed to be continuous, with chi square able to be any value. But when expected frequencies are small, the possible values of chi square are arithmetically quite limited, violating the assumption of continuity underlying the test. When 20 percent or more of the cells have expected frequencies less than 5, we report the significance as not applicable (abbreviated n.a.). There are alternative tests (e.g., Fisher's exact test) that may be used when expected frequencies are small. These alternatives are described in more advanced statistics texts listed in the Bibliography.

Since expected frequencies are functions of total row and total column frequencies (i.e., marginals), small expected frequencies are due to values of one or both variables having very few cases. This suggests solutions to the problem of small expected frequencies: Consider collapsing values so that the expected frequencies become 5 or more. Or, alternatively, consider excluding categories responsible for the small expected frequencies if the variable is nominal. These remedies for small expected frequencies are the same, of course, as those offered for small percentage bases in Section 5.7. As always, exclude or collapse frequencies only if it makes sense and helps you reach your research goals.

6.4   Statistical Significance Does Not Mean Substantive Significance

Another caution concerning the chi-square test (and, for that matter, all tests of statistical significance): *Statistical* significance does not necessarily mean *substantive* significance. We usually do not characterize relationships as substantively significant unless they are fairly strong. Who cares about a relationship showing just a few percentage points difference even if it is generalizable to the population (and thus is statistically significant)? Statistical significance depends not only on the strength of a relationship, but also on the number of cases in a sample. This makes intuitive sense since we have more confidence generalizing from larger samples than from smaller ones. But with enough cases, even a tiny, substantively trivial relationship will be statistically significant. Likewise, even a large percentage difference will not be statistically significant if it is based on very few cases.

Consider the three imaginary Tables 6.7a, b, and c describing the relationship between walking and quacking like a duck and actually being a duck. The hypothesis being tested is the old cliché: if it walks like a duck and quacks like a duck, it's a duck. Obviously we are interested in the numbers in these tables rather than the silly hypothesis, although they may help remind us that statistical significance is sometimes not quite what it is quacked up to be and we can't duck problems like small Ns or just walk away from them.

Table 6.7a.
Being a Duck by
Walking & Quacking
Like a Duck: 50 Cases
(in percentages)

Is a Duck?	Walks & Quacks Like Duck?	
	Yes	No
Yes	60	40
No	40	60
Total	100	100
(N)	(25)	(25)

$\chi^2 = 2.00$; n.s.

Table 6.7b.
Being a Duck by
Walking & Quacking
Like a Duck: 100 Cases
(in percentages)

Is a Duck?	Walks & Quacks Like Duck?	
	Yes	No
Yes	60	40
No	40	60
Total	100	100
(N)	(50)	(50)

$\chi^2 = 4.00$; $p < .05$

Table 6.7c.
Being a Duck by
Walking & Quacking
Like a Duck: 500 Cases
(in percentages)

Is a Duck?	Walks & Quacks Like Duck?	
	Yes	No
Yes	60	40
No	40	60
Total	100	100
(N)	(250)	(250)

$\chi^2 = 20.00$; $p < .001$

Each table shows the same 20 percentage point difference but with quite different significance levels, depending on the number of cases. With just 50 cases (Table 6.7a), the 20 percentage point difference is not statistically significant. With 100 cases (6.7b), the same 20 point difference is significant at the .05 level. And with 500 cases (6.7c), the same difference is significant at the .001 level. That is how it should be since we are more confident that a relationship "really" exists if it is based on more cases. With few cases, even a large difference cannot be generalized with confidence. And with enough cases, even an extremely small difference of no substantive significance can be generalized to the population.

When the N is very large, therefore, beware of researchers' boasts that relationships are "statistically significant." The boasts may well be true and yet the relationships may be trivially weak ones. Keep in mind that statistical significance only means that a relationship found in sample data can be confidently generalized to a larger population. That is very useful information about a relationship, but it does not by itself mean that a relationship is important, interesting, or worth attending to.

6.5   Significance Tests on Population Data

In this chapter, we have applied the chi-square test to sample data to find out how confidently we can generalize a relationship from a sample to the population from which the sample was drawn. This is a major application of inferential statistics like chi-square tests. We carry out tests of statistical significance to decide if we can generalize relationships found in sample data like the General Social Survey to the larger population. Significance tests allow us to rule out (with a given level of confidence) the possibility that our sample results are due to the randomness of sampling procedures.

But researchers often carry out tests of statistical significance even with population data such as data for the 50 most populous nations or the 50 American states or the 12 Canadian provinces and territories. True, there is no need to apply significance tests in order to generalize population data to the population. After all, if we have population data, then we already know about the population and have no need to generalize to it. With population data we are already 100 percent certain (i.e., $p = 1.00$) that the relationship found in the data occurs in the population. However, we often carry out significance tests anyway in order to assess the likelihood that a relationship found in population data is due to chance or random processes of any kind.

For example, if a chi-square test on a bivariate relationship found in population data indicates that the relationship is statistically significant at the .05 level, then we can be quite confident that the variables are not associated with one another simply by chance. It is unlikely that we would have gotten the bivariate table if cases were randomly distributed across cells. The relationship probably hasn't "just happened." Rather, in all likelihood there is some reason for the relationship other than random processes, and it is our job as scientific researchers to explain why the variables are associated.

I must caution you, however, that your instructor may think differently about the appropriateness of significance tests with population data. She/he may want to discuss this issue in your class.

6.6   Summing Up Chapter 6

Here is what we have learned in this chapter:

- Tests of statistical significance report the probability that we would find a relationship as strong as the one in our sample data if there is no relationship in the population from which the sample was drawn.

- In significance tests, the null hypothesis usually asserts that two variables are unrelated to one another in the population from which the sample was drawn.

- The .05 level of significance is usually used in significance tests. That is, when the probability of finding a relationship in sample data if the two variables are not related in the population is less than .05, we reject the null hypothesis that the two variables are unrelated.

- Rejection of a null hypothesis that is actually true is a Type I error. Failure to reject a null hypothesis that is actually false is a Type II error.

- The significance level is the probability of committing a Type I error.

- The chi-square test of significance is appropriate for cross-tabulations.

- The chi-square test compares observed frequencies with the frequencies we would expect if the two variables were unrelated.

- The chi-square test is inappropriate if 20 percent or more of cells have expected frequencies less than 5. In such situations, consider sensibly excluding or collapsing values with small frequencies to increase expected frequencies.

- Statistical significance depends on both the strength of a relationship and the number of cases in a sample.

- Statistical significance does not mean a relationship is substantively significant. Even a weak relationship will be statistically significant if based on enough cases.

- Tests of statistical significance are often used with population data to assess the likelihood that relationships are due to chance or random processes.

Key Concepts and Procedures

Ideas and Terms

test of statistical significance
null hypothesis
rejection of the null hypothesis
failure to reject the null hypothesis
level of statistical significance
probability
Type I error (alpha error)
alpha level

Type II error (beta error)
power
chi square
chi-square test
observed frequency
expected frequency
degrees of freedom (df)

Symbols

H_0
χ^2
f_e
f_o
df
$p < .05; p < .01;$ and $p < .001$
Prob. = n.a.

Expected frequency

Formulas

$$f_e = \left(\frac{\text{Row Marginal}}{N} \right) \text{Column Marginal}$$

$\longrightarrow X$

$$X^2 = \Sigma \frac{(f_o - f_e)^2}{f_e}$$

$$df = (r-1)(c-1)$$

$$\frac{(O.F. - E.F.)^2}{E.F.}$$

Add all up = Chi square

CHAPTER 7
Measures of Association for Cross-tabulations

We have already learned one measure of the strength of a relationship: percentage point differences across categories of the independent variable. The bigger the differences, the stronger a relationship. But percentage point differences are not a very concise measure, especially for larger tables that have many percentage differences to assess. It would be useful to have a single number giving an overall assessment of the strength of an association between two variables. Such a *measure of association* could both describe the strength of a single relationship such as Education → Civil Disobedience and allow comparisons of similar measures for other relationships.

We will learn several measures of association in this chapter. Measures with names like V and lambda, gamma, tau-c—and several others too. Each answers the second question asked at the start of Chapter 5: How strong is the relationship between two variables? Such measures also implicitly tell us if two variables are related. Furthermore, some measures of association also indicate the direction of a relationship if the variables are ordinal or interval/ratio.

After this chapter you will be able to:

1. Explain the purposes and conventions of measures of association.

2. Calculate and interpret common measures of association.

3. Recognize weak, moderate, and strong relationships using measures of association.

4. Understand and offer proportional reduction in error (PRE) interpretations of lambda and several ordinal measures of association.

5. Recognize the major advantages and disadvantages of C, V, ϕ, and the conditions under which each is appropriate.

6. Compare pairs of scores ordered in the same and different directions.

7. Recognize the major advantages and disadvantages of gamma, Somers' D, tau-b, and tau-c and the conditions under which each is appropriate.

8. Understand the difference between symmetric and asymmetric measures of association.

7.1   Overview of Measures of Association

Ideally and by convention, measures of association usually range either from 0 to 1.00 for nominal variables or from −1.00 through 0 to +1.00 for ordinal or interval/ratio variables. A value of 0 indicates no relationship between variables while a value of either −1.00 or +1.00 indicates a perfect relationship. Magnitudes (i.e., values disregarding the − and + signs) between 0 and 1.00 reflect the strength of a less-than-perfect relationship so that the stronger the relationship, the larger the magnitude of a measure of association. The sign (+ or −) of a measure of association for ordinal or interval/ratio variables indicates the direction of the relationship—positive or negative. The sign usually is omitted from positive relationships unless its omission might cause some ambiguity.

Statisticians have concocted dozens of measures of association, each with its own special interpretations and applications, advantages and disadvantages. In this chapter we will learn measures of association commonly used in the analysis of bivariate frequency and percentage tables. A major criterion for choosing one measure over another is the level of measurement of the variables in our table. Some measures of association are used if either or both of the variables are nominal. We will take them up first. Other measures are used if both variables in a table are rank-ordered (i.e., are either ordinal or interval/ratio). We will consider them later in the chapter.

7.2   Chi Square–Based Measures for Nominal Variables: C, V, and ϕ

C, V, and three closely related measures called lambda are measures of association used when either one or both variables are nominal. I first want to introduce C and V as well as a special case of V called ϕ, all measures based on chi square. Then I will describe lambda in the next section.

In discussing the chi-square test of statistical significance in Section 6.4, I pointed out that the magnitude of chi square depends on two factors: (1) the strength of the relationship and (2) the number of cases. Chi square is 0 if two variables are unrelated. That provides a good floor for a measure of association since, by convention, we want a measure of association to be 0 if variables are not related. Chi square has a maximum possible value of N times either r − 1 or c − 1, whichever is less. This means that chi square has a maximum of N for a 2 by 2 table, a maximum of 2N for a 3 by 3 table, and so on. If we could somehow adjust the value of chi square to eliminate the effect of the number of cases and take table size into account to adjust for a maximum value, then we would have a good measure of the strength of association. C and V are this kind of chi square–based measure.

British statistician Karl Pearson figured out one way to adjust chi square for the effect of the number of cases. Here is the formula for **Pearson's coefficient of contingency**, usually called simply C:

$$C = \sqrt{\frac{\chi^2}{\chi^2 + N}}$$

where C = contingency coefficient

χ^2 = chi square

N = total number of cases

The division of χ^2 by $\chi^2 + N$ adjusts for the number of cases.

Here's an example using C to measure the strength of the relationship between race and opinion about racial discrimination. Noting that Blacks have worse jobs, income, and housing than Whites, the General Social Survey asked respondents if these differences are mainly due to discrimination—Yes or No. Table 7.1 reports percentages by race.

Table 7.1. Opinion About Cause of Black Disadvantage by Race (in percentages)

Discrimination?	Race		
	White	Black	Other
Yes	34.6	64.2	53.1
No	65.4	35.8	46.9
Total	100.0	100.0	100.0
(N)	(1493)	(260)	(98)

$\chi^2 = 89.127$; df = 2; p > .001

C = .21

Not surprisingly, almost two-thirds of Blacks, but only a little over a third of whites, responded Yes, the situation of Blacks is mainly due to discrimination. With a chi square of 89.127 and 2 degrees of freedom, this relationship is statistically significant at the .001 level. We find C this way:

$$C = \sqrt{\frac{\chi^2}{\chi^2 + N}}$$

$$= \sqrt{\frac{89.127}{89.127 + 1851}}$$

$$= \sqrt{\frac{89.127}{1940.127}}$$

$$= \sqrt{.0459}$$

$$= .21$$

For a given N, the stronger the relationship, the larger the value of chi square and thus the larger the value of C. The C of .21 indicates a moderately strong relationship between race and opinion about discrimination as the main source of Blacks' disadvantages.

C has a lower limit of 0, and the stronger the relationship, the larger C is. That's fine. But C has a serious problem: Its upper limit depends on the number of rows and columns. It can never be larger than $\sqrt{\frac{\text{Min}(r-1, c-1)}{\text{Min}(r, c)}}$, where r is the number of rows, c is the number of columns, and Min(r − 1, c − 1) is either r − 1 or c − 1, whichever is less. (Min stands for *minimum*.) The maximum possible C for a table like Table 7.1 with two rows or two columns is only .71 (i.e., $\sqrt{\frac{1}{2}} = .71$). The upper limit of C for a table with three rows and three columns is only .82 (i.e., $\sqrt{\frac{2}{3}} = .82$). The upper limit increases as the minimum number of rows or columns increases (making C more appropriate for larger tables), but the upper limit is always less than 1.00. The lowered ceiling of C makes it somewhat difficult to interpret, rendering C less than an ideal measure of association. *Cramer's V*—often shortened to just *V*—is similar to C, but adjusts for the number of rows and columns so that it can reach 1.00 and its upper limit does not depend on the size of the table. Here is the formula for V:

$$V = \sqrt{\frac{\chi^2}{(N) \, \text{Min}(r-1, c-1)}}$$

where V = Cramer's V

χ^2 = chi square

N = total number of cases

r = number of rows

c = number of columns

Again, $\text{Min}(r-1, c-1)$ is either $r-1$ or $c-1$, whichever is less. $(N)\text{Min}(r-1, c-1)$ is the upper limit of χ^2 (allow me to skip the mathematical proof of this). Dividing by this expression limits the maximum value of V to 1.00.

Here is the calculation of V for the Race $\rightarrow$ Discrimination Opinion example of Table 7.1:

$$V = \sqrt{\frac{\chi^2}{(N) \, \text{Min}(r-1, c-1)}}$$

$$= \sqrt{\frac{89.127}{(1851) \, \text{Min}(2-1, 3-1)}}$$

$$= \sqrt{\frac{89.127}{1851}}$$

$$= \sqrt{.0482}$$

$$= .22$$

For a table like Table 7.1 with two rows and/or two columns, $\text{Min}(r-1, c-1) = 1$ and thus the formula for V simplifies $V = \sqrt{\frac{\chi^2}{N}}$.

In this special case of a 2 by c table or an r by 2 table, V is the same as a measure of association denoted ϕ (pronounced fē or fī). In other words, $\phi = \sqrt{\frac{\chi^2}{N}}$. ϕ is used for tables with two rows and/or two columns. ϕ has the unfortunate property of an upper limit greater than 1.00 for tables with more than two rows and two columns, but it works quite well for properly sized 2 by c tables or r by 2 tables. Researchers often square ϕ and report it as ϕ^2, called the *mean square contingency*. Since there are two rows in Table 7.1, rather than describing the strength of the relationship with Cramer's V, we would report that $\phi = .22$ or $\phi^2 = .05$.

As with other measures of association, V (or ϕ) = 0 indicates that there is no relationship between variables and V (or ϕ) = 1.00 indicates a perfect relationship. Values between 0 and 1.00 reflect the strength of the relationship. Again, a ϕ of .22 indicates that opinion about discrimination has a moderately strong relationship to race. Based as it is on chi square, ϕ can never be negative. That's good because relationships between nominal variables have no direction.

C and V (and V's special form, ϕ) are *symmetric measures of association*, which means that their values do not depend on which variable is the independent variable and which variable is the dependent variable. These measures of association are based on chi square, and chi square makes no distinction between independent and dependent variables. C and V (or ϕ) would be unchanged if we decided (very foolishly) to regard opinion about discrimination as the independent variable and race as the dependent variable.

Since C and V (or ϕ) are based on chi square, the chi-square test of statistical significance for the frequency table also applies to these measures of association. If the chi square for the table is statistically significant, so too is the chi square–based measure of association. If the chi square is not significant, neither is the measure of association. The assumption of random sampling necessary for the chi-square test also applies to the significance test for C or V (or ϕ).

7.3   Lambda

Guttman's coefficient of predictability, mercifully called just *lambda* and symbolized with the Greek λ, offers still another measure of association for nominal variables. Lambda measures the strength of a relationship by calculating the proportion by which we reduce errors predicting a dependent variable score if we know the value of the independent variable score for each case. Unlike C and V, lambda is not based on chi square and is asymmetric with regard to the independent and dependent variables. *Asymmetric* means that the value of lambda depends on which variable is dependent and which variable is independent.

Lambda is calculated from bivariate frequencies, not percentages. Here are frequencies for the Race → Discrimination Opinion example introduced in the previous section:

| | Race | | | |
Discrimination?	White	Black	Other	Total
Yes	516	☞ 167	☞ 52	735
No	☞ 977	93	46	☞ 1116

I have left off the total (N) row at the bottom because we do not need its information to calculate lambda. However, we need the marginal distribution for the dependent variable, so I have added a Total column for Discrimination Opinion. Since lambda makes use of modal frequencies, I have pointed out the mode within each column—Yes for Blacks and Other, No for Whites and the total sample.

In calculating lambda, we predict each case's score on the dependent variable—that is, each case's Discrimination Opinion. We first assume that we do not know independent variable (Race) scores, so let's focus on the Total distribution in the right-hand column.

Yes		735
No	☞	1116

Since No is the modal category for all 1851 respondents, we minimize our prediction errors if we place all 1851 cases into the No category. Each case is more likely to report No than Yes. In predicting No for each case, we will be correct 1116 times (the number of respondents who answered No). We will make 735 errors since that is how many cases do not report No. But that is the best we can do guessing in which dependent variable category each case belongs if we do not know anything else about cases (like their independent variable scores).

Now suppose we know the score of each case on our independent variable, Race. If we know that a respondent is White, we minimize our errors by placing that case and all other Whites into the No category, the modal response for Whites. We will make 516 errors since that many Whites did not answer No. Likewise, if we know a respondent is Black, we minimize our errors by placing that case and all other Blacks into the Yes category, the Blacks' modal response. In doing so, we will make 93 errors, but again, we can't do any better. Similarly, we place all 98 Other respondents into their modal category, Yes, and make 46 errors doing so. I trust that you see how those 655 errors were found. (Here's a clue: 516 + 93 + 46 = 655)

Now, let's see where we stand in counting errors. In our predictions of dependent variable scores using modes of independent variable categories, we make a total of 655 errors. Therefore, we will have reduced our errors from 735 to 655 if we know each case's score on the independent variable. Knowing cases' race reduced errors by 80 (i.e., 735 − 655 = 80).

As noted earlier, lambda is the proportion by which we reduce errors in predicting the dependent variable score if we know the independent variable score of each case. Here is lambda expressed as a formula:

$$\text{Lambda} = \frac{\text{Errors if we don't know IV Scores} - \text{Errors if we know IV Scores}}{\text{Errors if we don't know IV Scores}}$$

IV stands for independent variable. Applied to our Race → Discrimination Opinion example, we get:

$$\text{Lambda} = \frac{\text{Errors if we don't know IV Scores} - \text{Errors if we know IV Scores}}{\text{Errors if we don't know IV Scores}}$$

$$= \frac{(735 - 655)}{735}$$

$$= \frac{80}{735}$$

$$= .11$$

The numerator tells us how much we reduce errors if we know independent variable scores. The denominator is the total number of errors. Thus, lambda reports the proportion by which we reduce errors if we know independent variable scores. We reduce our errors in predicting discrimination opinion about 11 percent if we know each respondent's race. This lambda indicates a moderate relationship between the two variables.

The lambda computed above assumes that the column variable is the independent variable and the row variable is the dependent variable. That is how we have agreed to set up our tables while learning statistics. But we can also compute lambda treating the row variable as independent and the column variable as dependent. These two values of lambda will usually be different, sometimes remarkably so. (In the above example, lambda treating race as the dependent variable is .00 for reasons noted below.)

There is also a symmetric form of lambda, which is a kind of average of the two asymmetric lambdas. This symmetric lambda has some statistical limitations, however, and is not used often. Therefore, we have three lambdas for a given table, one measuring the strength of the effect of the column variable on the row variable, another measuring the strength of the effect of the row variable on the column variable, and the third a symmetric form of lambda. We will usually use the first of these lambdas since we usually set up our table with the independent variable in columns and the dependent variable in rows.

Unlike C or V, lambda is a *proportional reduction in error (PRE)* measure of association.[1] PRE measures of association have the important advantage of an intuitively meaningful interpretation. They tell us the proportion by which we reduce errors in predicting dependent variable scores if we know the score of each case on the independent variable. As we have seen, lambda in the Race → Discrimination Opinion example indicates that knowing the race of each respondent allows us to reduce errors in predicting discrimination opinion by 11 percent.

Lambda's PRE interpretation is very helpful. So too is lambda's possible variation between 0.00 and 1.00 for tables of any size. Moreover, lambda can never be greater than 1.00. Lambda has a significant disadvantage, however, in that it may produce a value of 0 even in situations in which there "really" is a relationship between variables. This occurs when one of the categories of the dependent variable is much larger than the others, so that in our lambda "predicting game" we decide to place all cases in that modal category no matter what the value of the independent variable. In other words, if each category has the same mode, lambda will be 0. (This situation occurs if Discrimination Opinion were treated as the independent variable in the preceding example. White would be the modal value for predictions from both the Yes and the No categories. We would make just as many errors knowing as not knowing independent variable scores, and lambda would be .00.) Moreover, even if a single dependent variable category does not completely dominate the others, it is likely that the smaller categories will not enter into the calculation of lambda at all.

The implication of this disadvantage: Lambda is a poor measure of association to use with a skewed dependent variable. Unfortunately, many social scientific variables are sufficiently skewed that lambda often fails to pick up on their relationships. Lambda, therefore, is not nearly as useful as we might hope from its PRE interpretation. Still, the ease of interpreting lambda as a PRE measure makes it a very desirable measure of association for relationships with less skewed dependent variables.

7.4 Choosing a Nominal Measure of Association

Let's take stock: If one or both variables in a bivariate cross-tabulation are nominal, we have four measures to choose among: C, V, ϕ (if the table is 2 by c or r by 2), and lambda. C has the serious problem of an

---

[1] We will find in Chapter 11 that ϕ^2 is also a PRE measure of association.

upper limit of less than 1.00. That makes C hard to interpret. Therefore, V (or φ) is usually preferable to C, especially for smaller tables. φ is used, of course, only for tables with two rows and/or two columns. Lambda both behaves well (i.e., varies between 0.00 and 1.00) and has a PRE interpretation—that's very good. However, lambda understates the strength of a relationship for a skewed dependent variable. With skewed dependent variables, we are usually better off using V or C. For dependent variables that are not skewed, we usually are better off with lambda.

But, as always, we need to *think* about what we are doing when we select a measure of association. Choice of a measure of association should be based on the research problem we are addressing, the nature of our data, and the characteristics of the various measures of association. A saw, hammer, and screwdriver are all useful tools. Which tool carpenters use depends on what they want to build and the materials they are working with. Carpenters don't use a screwdriver to pound a nail or a hammer to saw a board. The same is true for measures of association. Each has its uses, and a good researcher selects among measures of association very carefully. Creating elegant statistical tables is not unlike creating fine wooden tables. Both call for craftsmanship.

Although we typically report a single measure of association for a table, for purposes of comparison Tables 7.2, 7.3, and 7.4 present examples of C, φ, and lambda for each of the three tables we saw in Section 5.3. I report all three nominal measures of association in these tables to give you some sense of how they vary depending on the strength of a relationship.

Table 7.2.
Belief in an Afterlife by Gender (in percentages)

| Afterlife? | Gender | |
	Male	Female
Yes	82	83
No	18	17
Total	100	100
(N)	(759)	(978)

$\chi^2 = .471$; df = 1; n.s.
C = .02
φ = .02
Lambda = .00

Table 7.3.
Gun Ownership by Gender (in percentages)

| Own Gun? | Gender | |
	Male	Female
Yes	48	35
No	52	65
Total	100	100
(N)	(848)	(1066)

$\chi^2 = 32.7$; df = 1; p < .001
C = .13
φ = .13
Lambda = .00

Table 7.4.
Fear of Walking in Neighborhood by Gender (in percentages)

| Fear Walking? | Gender | |
	Male	Female
Yes	26	55
No	74	45
Total	100	100
(N)	(846)	(1057)

$\chi^2 = 169.6$; df = 1; p < .001
C = .29
φ = .30
Lambda = .14

You can also see that lambda fails to pick up even the moderately strong relationship of gun ownership to gender in Table 7.3. The reason is that gun ownership is skewed so that the modal category is No for both males and females. In Table 7.4 lambda picks up the relationship between gender and fear of walking in one's neighborhood because males and females have different modal responses.

7.5   Measures of Association for Ordinal Variables: Gamma

Now let's turn to four other measures of association: gamma, D_{yx}, tau-b, and tau-c. Each of these closely related statistics is appropriate for a cross-tabulation in which *both* variables are measured at either the ordinal level or the interval/ratio level.[2] All but tau-c are proportionate reduction in error (PRE) measures of association although their prediction rules are not as intuitively obvious as that for lambda.

Let's begin with Goodman and Kruskal's *gamma* (pronounced with a hard G, and symbolized by G for sample data and the Greek γ for population data). Like lambda, gamma is based on predictions of scores. But the predictions are made quite differently for gamma and take advantage of the ordinality of variables.

Here is how gamma works: For every pair of cases in a bivariate table, we predict that the rank-order of their scores on the dependent variable will be the same as the pair's rank-ordering of scores on the independent variable. If the variables are positively related, we expect cases with higher scores on the independent variable to also have higher scores on the dependent variable. After all, that's what it means to say variables are positively related: Higher scores on one variable are associated with higher scores on the other variable. Of course, unless the variables are perfectly related, expectations for some pairs will be wrong (i.e., we will make some errors), but our expectations will be right more often than wrong. The more strongly related the variables, the higher the proportion of times that we'll be right.

Consider, once again, the Education → Civil Disobedience relationship. We first saw this relationship in Chapter 5, and in Chapter 6 we carried out a chi-square test using the frequencies underlying these percentages. Both variables are ordinal. That's obvious for Education,

---

[2] One partial exception: For 2 by 2 tables, gamma is equivalent to Yule's Q, a measure of association defined only for a 2 by 2 table. If the sign of gamma is dropped, it may be interpreted as Q and used for nominal variables. Thus, although MicroCase does not compute Yule's Q as such, you can use gamma, dropping its sign and interpreting it as Q, for 2 by 2 tables involving nominal variables.

with ordered categories less than high school, high school, and college. But Civil Disobedience too is ordinal. It's values—Obey the Law and Follow One's Conscience—are ordered on a dimension that runs from extreme belief in always obeying the law to extreme belief in always following one's conscience. We can treat scores of Follow Conscience as higher than scores of Obey Law. Admittedly, treating Follow Conscience as higher than Obey Law is arbitrary, but that's no problem as long as we are consistent and keep the same ordering of values throughout our analysis. (Treating scores as "higher" or "lower" uses those adjectives in a technical, not moral or ethical, sense. No value judgment is implied.)

Treating Follow Conscience scores as higher than Obey Law scores, we found in Chapter 5 that Civil Disobedience is positively associated with Education. Therefore we expect—predict, if you will—that a respondent with more education will also give priority to following one's conscience even if it means breaking the law. That is, if respondent Joe College has more education than respondent Ed Hi-schooler, then we predict that Joe College will also score higher on Civil Disobedience than Ed Hi-schooler. The more strongly the variables are associated, the more such predictions will be true. Gamma is the proportion of all such comparisons of pairs of scores that fit this pattern.

Gamma sounds complicated, but its computation isn't difficult—just a little tedious. Here again are the bivariate frequencies for the 50 cases in our Education → Civil Disobedience example:

Civil Disobedience	Education		
	Less Than Hi School	High School	College
Conscience	4	13	12
Obey Law	6	11	4

We first saw these frequencies in Chapter 5. I have omitted both column and row totals because we do not need them to calculate gamma.

Let's begin in the lower left-hand corner. Note that each of the 6 respondents with less than a high school education who have Obey Law scores are "properly" rank-ordered with each of the 13 high school graduates with Follow Conscience scores. Since each of the 6 lower left-hand cases can be paired with each of the 13 High School/Follow Conscience cases, there are a total of (6)(13) = 78 pairs involving these cases.

Likewise, each of the 6 cases in the lower left-hand corner can be paired in the expected direction with each of the 12 college-educated respondents with Follow Conscience scores. That's another (6)(12) = 72 pairs with Education and Civil Disobedience ordered in the predicted direction.

And, too, the 11 high school graduates with Obey Law scores can be paired with the 12 College/Follow Conscience cases, giving another (11)(12) = 132 pairs ordered in the same direction on Education and Civil Disobedience. We can now stop looking for pairs ordered in the expected or predicted direction because there are no more cases that can be paired with other cases that have both more education and higher Civil Disobedience scores. If the table were larger, we would keep pairing cases in any cell with cases in cells above and to the right.

Thus, a total of 78 + 72 + 132 = 282 pairs are ordered in the same direction on both variables. For each of these 282 pairs, the case with a higher Education score also has a higher Civil Disobedience score.

To compute gamma, then, we first find the number of pairs having the same ranking on the two variables, Education and Civil Disobedience. To formalize what I said above: We multiply the frequency of each cell by the frequencies in the cells above and to the right of it, and then add all these products. Schematically in the diagram below, we multiply the frequency in each dark-shaded cell by the sum of frequencies in the light-shaded cells, and then add these products:

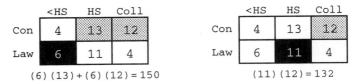

You have worked through the logic of this procedure in the preceding two paragraphs. Arithmetically:

$$\text{Same} = [(6)(13) + (6)(12)] + (11)(12)$$
$$= [78 + 72] + 132$$
$$= 282$$

There are 282 pairs of cases with scores ranked the same way on each variable.

Pairs having opposite rankings on the two values are counted analogously—that is, by multiplying each cell by the number of cases below and to the right of it, and then summing the products. These are pairs of cases in which the case with the higher independent variable score has the lower dependent variable score. For each pair, the case with more education believes in always obeying the law; the case with less education believes in following one's conscience. Schematically, multiply the frequency in the dark-shaded cell by the sum of the frequencies in the light-shaded cells, and then add the products:

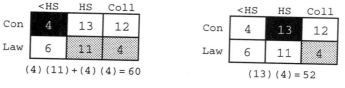

$$(4)(11)+(4)(4)=60 \qquad (13)(4)=52$$

Arithmetically:

$$\begin{aligned} \text{Opposite} &= [(4)(11) + (4)(4)] + (13)(4) \\ &= [44 + 16] + 52 \\ &= 60 + 52 \\ &= 112 \end{aligned}$$

So 112 pairs of cases are ordered such that the respondent with *more* education has a *lower* Civil Disobedience score.

Once we have the numbers of pairs ordered in the same and opposite directions, we calculate gamma with this formula:

$$\begin{aligned} G &= \frac{\text{Same} - \text{Opposite}}{\text{Same} + \text{Opposite}} \\ &= \frac{282 - 112}{282 + 112} \\ &= \frac{170}{394} \\ &= .43 \end{aligned}$$

Gamma = .43, indicating a fairly strong relationship between Education and Civil Disobedience. Notice that the numerator tells us how many more pairs are in the positive than the negative direction, while the denominator is the total number of pairs. Thus, a positive gamma reports the proportion by which we improve our predictions by "guessing" that the case with a higher independent variable score also has a higher dependent variable score. A negative gamma has an analogous interpretation except that the predict rule is that a case with the higher score on the independent variable has a lower score on the dependent variable. This is gamma's interpretation as a PRE measure of association.

A little tedious? Sure . . . but not hard. And computers take even the tedium out of computation of gamma.

The magnitude of gamma ranges from 0 to 1.00. If there is no relationship between variables, the number of pairs in the same direction equals the number of pairs in the opposite direction and gamma = 0.

A positive or negative sign tells the direction of the relationship. A positive relationship will always have more pairs in the same than in the opposite direction, and thus a positive gamma. Gamma will be negative, of course, if there are more pairs in the opposite than in the same direction. With values arranged from low to high across for the independent variable and up for the dependent variable, in the unlikely situation in which all cases lie along the major diagonal (from lower left to upper right) in a perfectly square table, all pairs are in the same direction and gamma equals 1.00—a perfect positive relationship. And, in a perfectly square table, if all cases lie along the minor diagonal (from upper left to lower right), gamma equals –1.00, a perfect negative relationship.

A crucial point: The "mechanics" and interpretation of computing gamma described above work fine as long as bivariate tables are arranged with independent variable values running in columns from left to right and dependent variable values running in rows from bottom to top. If either the independent or the dependent variable has values arranged differently, then an adjustment must be made in either computational procedures or interpretation of the sign of gamma. This caution also applies to interpretation of a gamma calculated by MicroCase and other computer programs. The direction of a computer-calculated gamma depends on the coding of variables. We could, of course, tinker with those codes to ensure that they run in "proper" directions so that the sign of gammas comes out "right." I usually find it easier, however, to let the computer calculate gamma for me, and then I assign the sign—positive or negative—depending on my reading of the percentage table. (These cautions and ways to handle the sign of gamma also apply to all gamma-related measures of association discussed in the remainder of this chapter.)

Gamma tends to run higher than other ordinal measures we will learn later in this chapter, especially for small tables. This is because gamma does not take into account ties on the dependent variable—that is, pairs of cases with the same dependent variable scores but different independent variable scores. Note that in computing gamma, we take into account only pairs of cases with different values on both the independent and dependent variables. No ties are involved.

Sometimes it is reasonable, however, to assume responsibility for predicting the ordering of dependent variable scores for tied pairs even if the predictions cannot be correct since the pairs are tied on the dependent variable. Consider, for example, pairs of cases, each with an

Obey Law score on Civil Disobedience, but with different scores on Education. The different Education scores would lead us to predict differences in Civil Disobedience. These predictions cannot be correct, however, since the pairs are tied on Civil Disobedience (both cases are Obey Law). We may nevertheless want to consider these ties in measuring the strength of the association, counting them in the total number of pairs. If there are many such ties (as there often are in cross-tabulations), gamma may be based on relatively few cases since it ignores ties. Gamma may run high. This is a serious shortcoming for gamma as a measure of association since gamma will then exaggerate the "real" strength of a relationship. For this reason, statisticians caution against using gamma for tables with many ties and turn to other measures of association that take ties into account. We will take up these alternatives to gamma in the rest of this chapter.

7.6   Somers' D_{YX}

Some other PRE measures of association for ordinal variables are designed to overcome this problem of ties by incorporating them into their computations. One of these measures is *Somers' D_{YX}* named for Robert H. Somers.[3] D_{YX} treats the row variable (called the Y variable) as dependent and the column variable (designated X) as the independent variable. Statisticians conventionally denote dependent variables as Y and independent variables as X, and then position the dependent variable symbol first in subscripts of measures of associations. We will see this notational system repeatedly through the rest of this text. Like gamma, D_{YX} is based on the number of pairs in the same and opposite directions. But unlike gamma, D_{YX} also incorporates the numbers of pairs that have identical scores on the dependent variable. That is, D_{YX} takes into account the number of pairs tied on the row (or Y) variable but not tied on the column (or X) variable.

I know this sounds a little like something devised by the Mad Hatter or those S.A.T. people, but concentrate hard and follow along. It's really quite easy. Let's take an example. Considering once again Education → Civil Disobedience, we'll follow convention by denoting the row variable (Civil Disobedience) as Y and the column variable (Education) as X. And let's use T_Y for the number of pairs tied on row variable Y but not on column variable X.

We find T_Y by multiplying the frequency in the dark-shaded cell by each frequency in the light-shaded cells, and then adding these products:

---

[3] Somers' D is often symbolized with a lowercase d, but MicroCase uses an uppercase D and that's the practice used throughout this text and the accompanying workbook.

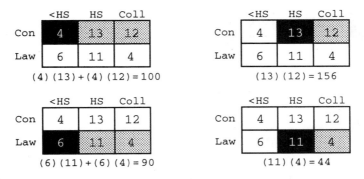

$$(4)(13) + (4)(12) = 100 \qquad (13)(12) = 156$$

$$(6)(11) + (6)(4) = 90 \qquad (11)(4) = 44$$

Thus:

$$T_Y = [(4)(13) + (4)(12)] + (13)(12) + [(6)(11) + (6)(4)] + (11)(4)$$

$$= [52 + 48] + 156 + [66 + 24] + 44$$

$$= 100 + 156 + 90 + 44$$

$$= 390$$

Look very closely at these tables. I want you to see that all cases in the shaded cells of each table are tied on the dependent variable. For example, all cases in the one darkly shaded and two lightly shaded cells of the upper left-hand table have Follow Conscience scores on Civil Disobedience. All cases in the shaded cells of the lower left-hand table have Obey Law scores. And so on.

Once we have calculated T_Y along with the numbers of pairs in the same and opposite directions, we can calculate D_{YX} with this formula:

$$D_{YX} = \frac{Same - Opposite}{Same + Opposite + T_Y}$$

For the above example:

$$D_{YX} = \frac{Same - Opposite}{Same + Opposite + T_Y}$$

$$= \frac{282 - 112}{282 + 112 + 390}$$

$$= \frac{170}{784}$$

$$= .22$$

So what does the D_{YX} of .22 mean? How do we interpret it? The implied positive sign on D_{YX} indicates that Education and Civil

Disobedience are positively related—the more education, the "higher" Civil Disobedience. The D_{YX} of .22 indicates a moderate relationship. D_{YX} can vary in magnitude between 0 and 1.00, and .22 falls in a moderate range. D_{YX} is a PRE measure of association, so we can also interpret the D_{YX} of .22 to mean that 22 percent more pairs of cases are ordered in the same direction than in the opposite direction on Education and Civil Disobedience.

Note that D_{YX} (.22) is considerably smaller than gamma (.43) for the same table. This is because D_{YX} takes dependent variable ties into account. Gamma doesn't. Formulas for D_{YX} and gamma have the same numerator (Same – Opposite). But since the denominator in the formula for D_{YX} adds in ties (T_Y) and, therefore, must be larger than the denominator of gamma (except in the remarkable situation of no ties at all), D_{YX} is almost always smaller than gamma. So, for a given table, D_{YX} can never be greater than gamma, and is almost always smaller—sometimes much smaller if a high proportion of pairs with different independent variables scores have identical dependent variable scores.

You need to know that there are actually two versions of Somers' D. We have just seen one version—D_{YX}. An analogous measure of association—D_{XY}—incorporates ties in the column (or X) variable rather than the row (Y) variable. The formula for D_{XY} is just like the formula for D_{YX} except that it incorporates ties on the column variable (denoted T_X) into its formula:

$$D_{XY} = \frac{\text{Same} - \text{Opposite}}{\text{Same} + \text{Opposite} + T_X}$$

T_X is found analogously to T_Y. In our example:

	<HS	HS	Coll		<HS	HS	Coll		<HS	HS	Coll
Con	4	13	12	Con	4	13	12	Con	4	13	12
Law	6	11	4	Law	6	11	4	Law	6	11	4

| (4)(6) = 24 | (13)(11) = 143 | (12)(4) = 48 |

Arithmetically:

$$T_X = (4)(6) + (13)(11) + (12)(4)$$
$$= 24 + 143 + 48$$
$$= 215$$

and

$$D_{XY} = \frac{\text{Same} - \text{Opposite}}{\text{Same} + \text{Opposite} + T_X}$$

$$= \frac{282 - 112}{282 + 112 + 215}$$

$$= \frac{170}{609}$$

$$= .28$$

D_{XY} is used rather than D_{YX} if we construct tables with the independent variable in rows and dependent variable in columns. However, if we follow this text's practice of having our independent variable in columns and dependent variable in rows, then D_{YX} is the appropriate Somers' D to use.

Not so incidentally, the implication of the preceding paragraphs is that Somers' D is an asymmetric measure of association. In the above example, $D_{YX} = .22$ and $D_{XY} = .28$. As with lambda, it matters which variable is independent and which variable is dependent. Which to use—D_{YX} or D_{XY}? That's an easy question. If the row variable is the dependent variable (as suggested throughout this text), then use D_{YX} since that is the measure that treats the row variable as dependent. If the column variable is the dependent variable, then use D_{XY}.

A final word about D: All the cautions about interpreting the sign of gamma apply to D_{YX} and D_{XY} as well. The sign depends on how tables are arranged, which in turn depends on the coding of variable values. As a practical matter, it is usually easiest to let computers calculate D, but to assign a positive or negative sign on the basis of your own reading of the percentage table. Certainly do not accept the computer-generated sign on D without first making sure that the coding of variables reflects your own definition of low and high values on each variable.

7.7 Tau-b and Tau-c

Finally, let's consider tau-b and tau-c, two cousins of gamma and D_{YX}.

Kendall's tau-b (often symbolized t_b for sample data and τ_b for population data)[4] shares characteristics of both gamma and the D_{YX}. Like gamma, tau-b is symmetric and has a PRE interpretation,

---

[4] Kendall's tau-b, named after Maurice G. Kendall, should not be confused with Goodman and Kruskal's τ_b, a measure appropriate for nominal data. You won't be confused by these taus in this text since MicroCase does not compute Goodman and Kruskal's τ_b and we won't consider it, but Goodman and Kruskal's measure does appear in the statistics and research literatures.

although not a very straightforward one. But like D_{YX}, tau-b takes ties into account in predicting the rank-ordering of pairs. Indeed, tau-b takes both kinds of ties into account and is a kind of average of D_{YX} and D_{XY} that can be calculated by taking the square root of the product of D_{YX} and D_{XY}. Here are two equivalent formulas for tau-b:

$$\text{tau-b} = \sqrt{D_{YX} D_{XY}} \quad or$$

$$\text{tau-b} = \frac{\text{Same} - \text{Opposite}}{\sqrt{(\text{Same} + \text{Opposite} + T_Y)(\text{Same} + \text{Opposite} + T_X)}}$$

In the same Education → Civil Disobedience example used for gamma and Somers' D:

$$\text{tau-b} = \sqrt{D_{YX} D_{XY}}$$

$$= \sqrt{(.22)(.28)}$$

$$= \sqrt{.0616}$$

$$= .25$$

or

$$\text{tau-b} = \frac{\text{Same} - \text{Opposite}}{\sqrt{(\text{Same} + \text{Opposite} + T_Y)(\text{Same} + \text{Opposite} + T_X)}}$$

$$= \frac{282 - 112}{\sqrt{(282 + 112 + 390)(282 + 112 + 215)}}$$

$$= \frac{170}{\sqrt{(784)(609)}}$$

$$= \frac{170}{\sqrt{477,456}}$$

$$= \frac{170}{690.98}$$

$$= .25$$

Tau-b will be 1.00 or –1.00 only if its numerator and denominator are the same. Carefully consider the formula for tau-b and you will see that the denominator will equal the numerator only if there are no ties. Thus, tau-b cannot reach 1.00 in tables (like the one here) that are not square because nonsquare tables must have some ties. This is a potentially serious disadvantage of tau-b. So too is the complexity of its PRE interpretation.

Tau-c (sometimes symbolized t_c or τ_c for sample and population data, respectively) overcomes this problem for nonsquare tables. Tau-c's interpretation is similar to that of tau-b, although tau-c is not a PRE measure. However, tau-c can reach a maximum magnitude of 1.00 regardless of the shape of a table. The formula for tau-c is:

$$\text{tau-c} = \frac{2 \, \text{Min}(r, c) \, (\text{Same} - \text{Opposite})}{N^2 \, \text{Min}(r - 1, c - 1)}$$

where N = total number of cases

   r = number of rows

   c = number of columns

The number of same and opposite pairs is computed just as for gamma and Somers' D.

For the Education → Civil Disobedience example:

$$\text{tau-c} = \frac{2 \, \text{Min}(r, c) \, (\text{Same} - \text{Opposite})}{N^2 \, \text{Min}(r - 1, c - 1)}$$

$$= \frac{2 \, (2) \, (282 - 112)}{(50)^2 \, (1)}$$

$$= \frac{680}{2500}$$

$$= .27$$

The same cautions and ways to handle signs of gamma and Somers' D also apply to tau-b and tau-c. Signs depend on arrangement of tables, which depend on coding of variables, so be sure that the sign of an ordinal measure of association reflects your own definition of low and high values of variables.

7.8   Measures of Association: An Overview

In this chapter we have learned many measures of association, each with its own advantages and disadvantages and peculiarities. Table 7.5 presents an overview of their major characteristics.

Table 7.5 Summary of Nominal and Ordinal Measures of Association

Measure of Association	Appropriate Levels of Measurement	Range	Symmetric?	PRE Measure?	Additional Information
C	One or both variables are nominal.	0 to <1.00	Yes	No	Upper limit increases with table size, but is always less than 1.00.
Cramer's V	One or both variables are nominal.	0 to 1.00	Yes	No	Can reach +1.00 for table of any size. $V = \phi$ for 2 by c or r by 2 table.
ϕ	One or both variables are nominal.	0 to 1.00	Yes	Yes (ϕ^2)	Use only for 2 by c or r by 2 table. $\phi^2 = r^2$ for 2 by 2 table.
Lambda	One or both variables are nominal.	0 to 1.00	No	Yes	May underestimate strength if dependent variable is skewed. Symmetric lambda is also available.
Gamma	Both variables are ordinal.	−1.00 to 1.00	Yes	Yes	Tends to run high. Does not take ties into account.
Somers' D_{YX} (and D_{XY})	Both variables are ordinal.	−1.00 to 1.00	No	Yes	Use D_{YX} if dependent variable is the row variable.
Tau-b	Both variables are ordinal.	−1.00 to 1.00 (square tables)	Yes	Yes	Can reach ±1.00 only for square tables.
Tau-c	Both variables are ordinal.	−1.00 to 1.00	Yes	No	Can reach ±1.00 for table of any size.

Later on we will learn two additional measures of association appropriate for bivariate analysis of interval/ratio variables. In Chapter 9 we will encounter eta-squared (E^2), useful with a nominal or ordinal independent variable and an interval/ratio dependent variable. In Chapter 10 we will meet another measure of association, r (the correlation coefficient), that describes the strength of relationship between two interval/ratio variables. As Table 7.5 notes, ϕ^2 is equivalent to r^2. More about this equivalence three chapters from now.

How should you interpret the magnitudes of the nominal and ordinal measures of association we have learned in this chapter? How large should V or lambda or whatever be to say a relationship is strong? I wish I could give you some hard and fast rules. Alas, there are none. Interpretation of measures of association depends on the particular measure. Chi square–based measures like C and V tend to run low. We shouldn't be surprised that the race-discrimination relationship in Table 7.1 has a C of "only" .21 and V of "only" .22 despite a considerable percentage difference (almost 30 percentage points) across independent variable categories. In contrast, gamma tends to run high, so we would often want a larger gamma than, say, tau-c to conclude that a relationship is strong. Assessment of measures of association also depends on the distributions of the two variables. We have seen, for example, that a skewed dependent variable will often produce a lambda of 0 and thus fail to pick up whatever relationship there may be between two variables. And, too, the context of an analysis matters, including what previous research has found and what strengths of relationships are expected in a particular research tradition.

As a beginner, you will have to live with this ambiguity, confident that your skills in interpreting measures of association will improve with experience. You will start to acquire that experience with the workbook exercises accompanying this chapter.

7.9   Summing Up Chapter 7

Here is what we have learned in this chapter:

- Measures of association are numbers that summarize the strength of a relationship.

- Measures of association usually range in magnitude from 0 for no relationship to ±1.00 for a perfect relationship. The sign indicates the direction of the relationship for ordinal and interval/ratio variables.

- Choice of a measure of association depends in part on the levels of measurement of the variables.

- Proportionate reduction in error (PRE) measures of association are conveniently interpreted as the proportion by which knowledge of the independent variable reduces errors in predicting the dependent variable.

- C and V are symmetric chi-square–based measures of association for nominal variables.

- C's upper limit increases as the size of the table increases, but is always less than 1.00.

- V reduces to and is read as ϕ for tables with two rows and/or two columns.

- Lambda is a PRE measure of association for nominal variables based on assignment of cases to modal values of the dependent variable.

- Lambda underestimates the strength of a relationship for highly skewed variables.

- Gamma, Somers' D, and Kendall's tau-b are PRE measures of association for ordinal variables. All are based on predictions of the ordering of pairs of cases on the two variables in a bivariate table.

- Gamma is a symmetric measure that does not take into account ties on the dependent variable and tends to run higher than other ordinal measures.

- D_{YX} and D_{XY} are asymmetric measures that take into account ties on the row and column variables, respectively.

- Tau-b is a symmetric measure that takes into account ties on both the row and column variables. It can reach 1.00 only in square tables.

- Tau-c is a symmetric non-PRE measure of association for ordinal variables that can reach 1.00 regardless of the shape of a table.

Key Concepts and Procedures

Ideas and Terms

measure of association
chi square–based measure of association
C (Pearson's coefficient of contingency)
V (Cramer's V)
ϕ and ϕ^2
lambda
symmetric measure of association
asymmetric measure of association

PRE (proportional reduction in error)
  measure of association
gamma
same pairs, opposite pairs, and tied
  pairs
Somers' D_{YX} and D_{XY}
tau-b
tau-c

Symbols

C
V
ϕ and ϕ^2
Min(r, c)
Min(r – 1, c – 1)
G
λ
γ
D_{YX} and D_{XY}
tau-b or τ_b
tau-c or τ_c

Formulas

$$C = \sqrt{\frac{\chi^2}{\chi^2 + N}}$$

$$V = \sqrt{\frac{\chi^2}{(N)\,\text{Min}(r-1,\,c-1)}}$$

$$\phi = \sqrt{\frac{\chi^2}{N}}$$

$$\text{Lambda} = \frac{\text{Errors if we don't know IV Scores} - \text{Errors if we know IV Scores}}{\text{Errors if we don't know IV Scores}}$$

$$G = \frac{\text{Same} - \text{Opposite}}{\text{Same} + \text{Opposite}}$$

$$D_{YX} = \frac{Same - Opposite}{Same + Opposite + T_Y}$$

$$tau\text{-}b = \sqrt{D_{XY} D_{YX}} \quad and$$

$$tau\text{-}b = \frac{Same - Opposite}{\sqrt{(Same + Opposite + T_Y)(Same + Opposite + T_X)}}$$

$$tau\text{-}c = \frac{2\,Min(r, c)\,(Same - Opposite)}{N^2\,Min(r - 1, c - 1)}$$

ANALYSIS WRITE-UP 3: BIVARIATE CROSS-TABULATIONS

✦   ✦   ✦   ✦   ✦

A. *Here's an example of a write-up of a single bivariate cross-tabulation:*

General Social Survey respondents were asked how strongly they agreed or disagreed that employers should make special efforts to hire and promote qualified women. I collapsed the five-point Likert scale into three categories: agree, neither agree nor disagree, and disagree. Political party preference, reported on a seven-point scale ranging from strong Democrat to strong Republican, was collapsed into Democrat, Independent, and Republican categories.

Table 1 presents a cross-tabulation of support for affirmative action for women by political party preference. Democrats and Independents support affirmative action for women somewhat more strongly than do Republicans. On this issue of affirmative action for women, the close similarity of the Independents' distribution to that of Democrats is especially striking. Well over 60 percent of Democrats and Independents but only 42 percent of Republicans voice support for affirmative action by employers. This moderate relationship has a gamma of .32. With a chi square of 71.678 and 4 degrees of freedom, the relationship is statistically significant at the .001 level.

TABLE 1 ABOUT HERE

Treating political preference as an ordinal variable, I have chosen to use gamma as a measure of association. Gamma "picks up" the relationship in the table quite well. D_{YX} or tau-b would also be acceptable measures of association for this table.

Include the table at the end of the paper or report:

Table 1. Affirmative Action for Women by
Political Party Preference (in percentages)

Affirmative Action for Women	Political Party Preference		
	Democrat	Independent	Republican
Agree	64.2	61.4	42.0
Neither	12.7	15.5	15.1
Disagree	23.1	23.2	42.9
Total	100.0	100.1	100.0
(N)	(685)	(207)	(517)

$\chi^2 = 71.678$; df = 4; p < .001; G = .32

Alternatively, the following stacked bar graph could be substituted for Table 1:

Figure 1. Opinion About Affirmative Action for Women by
Political Party Preference

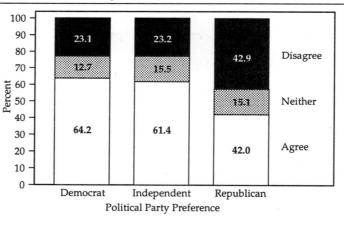

Because the table and stacked bar graph present nearly identical information, do not present both—choose one format or the other. If the stacked bar graph is used, the chi-square test should be reported in the text of the write-up.

B. *With a set of several related dependent variables, a single condensed table offers a more efficient way to present bivariate relationships. Table 2 describes gender differences on five independent variables concerning the right to suicide—all in one table.*

The General Social Survey asked respondents whether they think a person should have the right to take his/her own life under various conditions. Table 2 reports the overall percentages as well as the percentages of males and females agreeing that a person should have such a right. Although almost two-thirds of respondents support the right of a person with an incurable disease to take his/her own life, only small proportions support such a right for a person facing bankruptcy, dishonoring his/her family, or simply tired of living. Gender comparisons consistently indicate that males are more likely than females to support the right to take one's own life, although gender differences are quite small. The largest male-female difference is for a person with an incurable disease—about 70 percent for males and 60 percent for females. Gender differences are, however, statistically significant at the .001 level.

TABLE 2 ABOUT HERE

Include the table at the end of the paper or report:

Table 2. Opinions About Suicide by Gender (in percentages)

OK to Take Own Life If a Person . . .	All Cases	Gender		X^2	$p <$	V
		Male	Female			
Has incurable disease	64.4	69.5	60.4	16.709	.001	.10
(N)	(1859)	(814)	(1045)			
Has gone bankrupt	9.6	12.2	7.7	10.973	.001	.08
(N)	(1922)	(839)	(1083)			
Dishonored family	10.1	12.6	8.2	10.181	.001	.07
(N)	(1922)	(843)	(1079)			
Is tired of living	18.0	20.8	15.7	8.163	.001	.07
(N)	(1061)	(836)	(1061)			

CHAPTER 8
Comparison of Means and *t* Test

Some wag observed that there are two kinds of people: those who classify people and those who don't. I've long thought that the two kinds of people are those who find that observation amusing and those who don't. Anyway, dichotomies are everywhere: male/female, Black/White, native-born/foreign-born, married/unmarried . . . . So many variables that scientists (social/nonsocial) use are dichotomous. There are even two types of variables: dichotomous and non-dichotomous.

Often we use dichotomous independent variables with interval/ratio dependent variables. We compare incomes of males and females, years of education of African Americans and whites, family sizes of native and foreign-born citizens, number of friends of married and unmarried persons. In principle we can analyze such relationships with bivariate percentage tables—but only in principle, rarely in practice. Interval/ratio variables typically have far too many values to allow for manageable frequency and percentage tables. Of course, we can collapse variables with many values into a small, manageable number of values. But collapsing variables always loses information about differences among individual scores. With a variable such as years of education, for example, we lose the distinctions among respondents with 13, 14, 15, and 16 years of education by collapsing them into the same "college" category, thus effectively lowering the level of measurement from interval/ratio to ordinal. Moreover, cross-tabulation does not take advantage of interval/ratio variables' standard units of measurement. A standard unit of measurement allows mathematical operations like addition and division that in turn allow us to compute helpful statistics like means and standard deviations. Cross-tabulation is also relatively inefficient, requiring fairly large numbers of cases within independent variable categories for stable percentages.

So why work with categorized, "fuzzy" ordinal data when we have more precise interval/ratio data to begin with? Why use a statistical technique that does not take full advantage of the properties of interval/ratio dependent variables? Surely we can avoid this information loss. Indeed we can. With a dichotomized independent variable and an interval/ratio dependent variable, we can carry out a *difference of means test* that makes good use of the interval/ratio scale of the dependent variable. This difference of means test, often called a *t test*, is what this chapter is all about.

After this chapter you will be able to:

1. Identify situations in which you can compare means and apply *t* tests.

2. Interpret a box-and-whisker diagram for a dichotomous independent variable.

3. Understand the purpose of and carry out *t* tests of the differences between means.

4. Construct and interpret confidence intervals around differences between means.

5. Exercise caution in interpreting results of *t* tests when Ns are large.

6. Distinguish between and recognize the appropriateness of one-tailed and two-tailed tests of statistical significance.

8.1   Box-and-Whisker Diagrams/Differences Between Means

You have probably spent more hours watching television than you have studying in classrooms. I know I have . . . and with little apology! TV permeates contemporary life as it informs and misinforms, entertains and lulls and startles all of us. But like just about every cultural pattern, television watching varies across the social structure. Some people watch more television than others, and variations in television watching are related to location in the social structure. Consider gender differences in television watching. The General Social Survey asks respondents how many hours, on average, they watch TV each day. We could compare men's and women's TV watching using a bivariate table, but it is certainly more efficient to compare means. That is, we can compare the mean hours of TV watching for males with the mean hours of TV watching for females.

Here are the actual means from GSS data: 2.75 for men and 3.01 for women. So women on average watch TV a bit more than men. We find the actual gender difference by subtracting the female mean from the male mean: 2.75 – 3.01 = –.26. On average, men watch television about .26 hour a day less than women. That's about a 15-minute difference. (By the way, in finding male and female means I excluded outliers who report watching TV over 16 hours a day. Let me note, too, that whether we subtract the female mean from the male mean, or vice versa, is usually arbitrary and doesn't matter statistically.)

Figure 8.1 displays the male-female TV watching data in a special graph called, with some whimsy, a ***box-and-whisker diagram*** (sometimes more staidly called a ***boxplot***). The vertical arrays of dots show the distributions of dependent variable (TV watching) scores within each category of the independent variable (Gender) listed along the horizontal axis. The dots are better called data points because that's what they are. The height of each data point corresponds to a case's score on the dependent variable represented by the vertical axis. Many GSS respondents watch TV the same number of hours, of course, and so their data points are superimposed on one another. For example, 106 men report watching television 4 hours a day. Thus the dot for men who watch TV 4 hours a day actually represents 106 cases. There are really 106 data points stacked so neatly on top of one another that they look like a single data point. Even with this ambiguity in dots, the vertical arrays of data points give some sense of the distribution of scores within the dependent variable categories.

Figure 8.1. Box-and-Whisker Diagram for Daily Television Watching by Gender

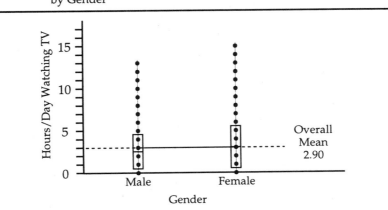

The horizontal lines within boxes describe category means—2.75 for males and 3.01 for females. Each elongated box in Figure 8.1 extends one standard deviation above and one standard deviation below the mean dependent variable score for that value of the independent variable. The male standard deviation is 2.03, so the box for males extends 2.03 standard deviations below the male mean. The box for females extends 2.2 hours—the female standard deviation—above and below the female mean. Thus, the longer the box, the greater the variance in dependent variable scores within that category of the independent variable.

Data points extend out from the standard deviation boxes. These are the whiskers of the box-and-whisker diagram. Data points in this example extend much further above than below both the male and female boxes because TV watching is skewed toward higher values for both genders. This sort of pattern is common for variables that have a "floor" or "ceiling." No one can watch television less than 0 hours a day.

For nominal independent variables like gender, we cannot attach any meaning to the solid line connecting box midpoints. But for ordinal or interval/ratio variables, the solid line connecting boxes indicates the direction of the relationship between the two variables—positive, negative, or curvilinear. We will consider interpretations of this line in the next chapter when we extend box-and-whisker diagrams to independent variables with more than two values.

Figure 8.1 also includes a dotted line indicating the overall TV watching mean—2.90—for all respondents (i.e., both males and females). This line gives us a reference for interpreting means of independent variable categories. We see here that the male mean is slightly below and the female mean is just a bit above the overall mean.

A box-and-whisker diagram gives us a visual display of means and the difference between them. Here is a generalization of this comparison of means: If our independent variable is dichotomous and our dependent variable is interval/ratio, we calculate and compare mean dependent variable scores for the two values of the independent variable. Working with means is analogous to working with percentages in bivariate percentage tables. Just as we calculate percentages within categories of the independent variable, so too we calculate means within categories of the independent variable. And just as we compare percentages across independent variable categories, so too we compare means across independent variable categories. Procedures for handling means and percentages are based on the same underlying logic. No problem.

Well, maybe one problem, the solution to which will occupy us the rest of this chapter. We need some way to confidently generalize the difference between sample means to the larger population from which the sample was drawn. Certainly we are not interested in the particular sample of Americans who happened, through random sampling, to end up in the General Social Survey. No, we want to say something about gender differences in TV watching among all 200 million or so American adults.

This is, once again and not for the last time, the fourth of the six questions about relationships posed at the beginning of Chapter 5. (Go ahead—look back at those six questions.) Just as the chi-square test gives us a basis for deciding if a relationship found in a bivariate percentage table can be generalized to the population, so too we need a test to decide if we can generalize a difference between means in sample data to the population. For example, does a gender difference in TV watching in the GSS data indicate that there is a gender difference in the whole U.S. adult population from which the GSS sample is drawn? We'll find out how to answer this question in the next section.

8.2   *t* Test for the Difference Between Means

Let's first get the "big picture" of generalizing a difference of means found in sample data to the population. When we compare dependent variable means (e.g., TV watching means) across independent variable categories (males and females), we are using data from only one sample (the particular GSS sample). But because sampling is random, the particular sample we have is only one of a huge number of possible samples that could have been drawn. Think about all those possible samples (one of which is our actual sample). Imagine finding the gender difference for the first sample, then the gender difference for the second sample, and then the third sample, and then the fourth sample . . . and on and on. We would eventually have a gigantic number of differences between means—zillions of differences between means, one for each of the zillions of possible samples. We could then form a distribution of these zillions of differences. That distribution would be a sampling distribution—that is, it would be the distribution of a statistic (here, the male mean minus the female mean) for all possible samples.

Now do a "what if" about these differences between means for the zillions of samples: What if there were no difference between means in the population? What would the sampling distribution of the difference of means be like? (Think hard about this. I'll wait for you in the next paragraph.)

Answer: If there were no difference between means in the whole population, then most of the samples would have differences between means that are quite small—somewhere around 0. Of course, just by chance, some of the samples would have large positive differences (i.e., the male mean would be much larger than the female mean). These would be samples with high proportions of male couch-potatoes and low proportions of TV-addicted females. And some samples would have large negative differences (i.e., few TV-addicted men and many TV-saturated women). But those would be fairly rare samples. For most samples, differences between means for males and females will be around zero if there is no male-female difference in the population.

On the other hand, what if there were a *large* difference between male and female means in the population? What if women watch either much more or much less television than men? Then we would find that most of the differences between sample means would be large (in either a negative or positive direction). If there is a large gender difference in the population, then we expect most samples to have a large gender difference and few samples would have male-female differences around zero.

To summarize thus far: The smaller the difference between means in the population, the greater the proportion of samples with small differences between means. The larger the population difference between means, the greater the proportion of samples with large differences between means. Yes, that makes sense.

Now let's proceed much as we did with the chi-square test in Chapter 6. Let's set up a null hypothesis that asserts that two variables are not related in the population. In our example, the null hypothesis asserts that there is no difference between the male and female means in the population. Symbolically, the null hypothesis is H_0: $\mu_1 = \mu_2$, where μ_1 is the TV watching mean for males and μ_2 is the TV watching mean for female respondents. Still another, identical way to state the null hypothesis: H_0: $\mu_1 - \mu_2 = 0$. If we can reject the null hypothesis that $\mu_1 - \mu_2 = 0$, then we can conclude that the variables are indeed related in the population—in other words, that men and women differ in their amounts of daily television watching.

We decide whether or not to reject the null hypothesis by finding where our actual sample's difference between means lies in the sampling distribution for all possible differences, assuming that $\mu_1 - \mu_2 = 0$. The location of the sample's difference in that sampling distribution tells us the probability that there is no difference in the population. If the difference between means in our sample is a fairly likely occurrence even if the variables are unrelated in the population, then we do not reject the null hypothesis and we conclude that the variables are not related. On the other hand, if the difference between means in our sample is a rare occurrence if the variables are unrelated in the population, then we reject the null hypothesis of no difference between means and conclude that the variables are related.

Now although we can imagine zillions of samples (each with a difference between means), we can't actually draw that many. Life is way too short. What we need is some statistic based on differences between means that has a known sampling distribution (i.e., it associates a probability with each possible sample outcome). We can then rely on that mathematically derived sampling distribution to decide whether we reject or do not reject the null hypothesis of no relationship in the population.

It turns out that the sampling distribution for the difference between means gets closer and closer to a normal distribution as the number of cases gets larger. Indeed, it becomes quite close to normal for even 50 or so cases. We could, therefore, get by pretty well by treating the sampling distribution as normal, at least for samples of 50 or more cases. There is no need to do so, however, because statisticians have derived a statistic that describes the sampling distribution even more closely.

The statistic we use is called **Student's *t*,**[1] or, more commonly, just *t*. Here is its formula (although I will soon simplify it):

$$t = \frac{(\overline{X}_1 - \overline{X}_2) - (\mu_1 - \mu_2)}{s_{\overline{X}_1 - \overline{X}_2}}$$

where $\overline{X}_1$ = dependent variable mean for category 1 of the independent variable for the sample

$\overline{X}_2$ = dependent variable mean for category 2 of the independent variable for the sample

μ_1 = dependent variable mean for category 1 of the independent variable for the population

μ_2 = dependent variable mean for category 2 of the independent variable for the population

$s_{\overline{X}_1 - \overline{X}_2}$ = standard error of the difference between means

The numerator of *t* is the difference between the sample difference of means ($\overline{X}_1 - \overline{X}_2$) and the population difference of means ($\mu_1 - \mu_2$). But since our null hypothesis usually asserts that there is no difference between the means of the two independent variable categories in the population, $\mu_1 = \mu_2$. Thus ($\mu_1 - \mu_2$) = 0, and that expression drops out of the numerator, simplifying *t* to:[2]

---

[1] Student's *t* is named after "Student," the pen name used by William S. Gossett when he first wrote on the statistic in a 1908 journal article. Gossett's employer, Dublin's Guinness Brewery, would not allow employees to publish their research, so Gossett obscured his identity behind "Student." Surely statistics has never served a higher purpose than the brewing of Guinness.

[2] In atypical cases in which we hypothesize a specific nonzero difference between means, we retain ($\mu_1 - \mu_2$) in the formula for *t*. Usually, though, we hypothesize no difference and, in effect, drop ($\mu_1 - \mu_2$).

$$t = \frac{(\overline{X}_1 - \overline{X}_2)}{s_{\overline{X}_1 - \overline{X}_2}}$$

The numerator of t is simply the difference between the dependent variable means for the two categories of the dichotomous independent variable. The denominator is the standard error of the difference between means. Ordinarily we do not know for sure what this standard error is, but we estimate it with this formula:

$$s_{\overline{X}_1 - \overline{X}_2} = \sqrt{\left(\frac{N_1 s_1^2 + N_2 s_2^2}{N_1 + N_2 - 2}\right)\left(\frac{N_1 + N_2}{N_1 N_2}\right)}$$

where s_1^2 = dependent variable variance for category 1 of the independent variable

   s_2^2 = dependent variable variance for category 2 of the independent variable

   N_1 = number of cases in category 1 of the independent variable

   N_2 = number of cases in category 2 of the independent variable

I agree that this is a complicated looking formula, but let's not be intimidated. The first parenthetical expression under the square root sign is a weighted average of the variances (s_1^2 and s_2^2) for each category of the dichotomous independent variable. The second parenthetical expression corrects for bias in that weighted average. The result of these mathematical operations is an unbiased estimate of the standard error of the sampling distribution of the difference between means. Note that the standard error is a function only of Ns and standard deviations, and does not depend on the means.

Now back to the formula for t: $t = \frac{(\overline{X}_1 - \overline{X}_2)}{s_{\overline{X}_1 - \overline{X}_2}}$. By dividing the difference between means in our sample data by the standard error, we are expressing that difference in terms of standard error units. The size of t thus tells us how many standard errors our difference is from 0. (Recall that a standard error is the standard deviation of a sampling distribution.) Notice that t may be either positive or negative, depending on whether $\overline{X}_1 - \overline{X}_2$ is positive or negative. Of course, t can also be 0 if $\overline{X}_1 - \overline{X}_2 = 0$, but in that case there is no difference between means in our sample and we would certainly not bother testing a null hypothesis of no difference. There is no way we could reject the null hypothesis of no difference between means in the population if our sample data shows no difference.

The t statistic is useful because statisticians have figured out the probability of getting each possible t. These probabilities are summarized in a table of *t* values in Appendix Table 2. Like chi square, the probability associated with a given *t* depends on degrees of freedom. Degrees of freedom for *t* is given by the simple formula

$$df = N_1 + N_2 - 2$$

where N_1 = number of cases in category 1 of the independent variable

N_2 = number of cases in category 2 of the independent variable

We'll find *t* and its associated probability for the gender → TV watching relationship shortly. First, however, I want to point out that the magnitude of *t* depends not only on the actual difference between means, but also on the number of cases and on the sizes of the standard deviations of the dependent variable within independent variable categories. For given standard deviations, larger Ns produce smaller standard errors. For given Ns, smaller standard deviations produce smaller standard errors.

This is as it should be. The more cases, the more confidence we rightly have in generalizing from a sample to the population. This makes good intuitive sense, and a little arithmetic demonstrates the impact of numbers of cases on the standard error. For convenience, let $s_1 = s_2 = 1$. Then compute the standard error first for $N_1 = N_2 = 10$ and then for $N_1 = N_2 = 40$. I'll let you do the math on this. If you do it right, you'll find standard errors of .47 for the smaller Ns and .23 for the larger Ns. As N increases, the standard error shrinks, reflecting the greater homogeneity of sample results.

It is a little harder to see why smaller standard deviations shrink the standard error. But compare the two sets of distributions in Figure 8.2. Each set has the same pair of means for males and females. The distributions on the left, however, have larger standard deviations than the distributions on the right. The wide spread of male scores in Sample A gives us relatively little confidence about the population mean for males. Our best guess is that it is near the male sample mean, but we can't be very sure—the sample scores are spread quite widely. Likewise, the relatively wide spread of female scores in Sample A gives us little assurance that the female population mean is near the female mean for the sample. So maybe Sample A is drawn from a population in which the male mean is higher than the male sample mean and the female mean is lower than the female sample mean. There is some reasonable chance, in fact, that the population from which the males and females in Sample A were selected has little or no difference between male and female means. In Sample B, with smaller standard deviations, there is little overlap between the male and female distributions. Since the male scores cluster closely together, we would expect that the population mean for males is close to the male mean for the sample. Likewise, the tightly clustered female scores give us some confidence that the female population mean is close to the sample mean for females. We would surely feel more confident generalizing from a difference between means from Sample B than from Sample A.

Figure 8.2. Pairs of Samples with Same Means and Different Standard
Deviations

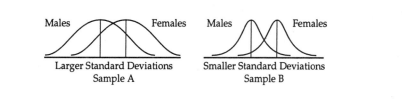

So smaller standard deviations give us greater confidence in generalizing to a population. This greater confidence is reflected by—and, indeed, is based on—the smaller standard errors associated with smaller standard deviations.

But now let's actually find t and decide whether the GSS data's male-female difference in TV watching can be generalized to the population. I reported above that the male and female means are 2.75 and 3.01, with a difference of –.26. Here are also the standard deviations and Ns that we need to calculate $s_{\bar{X}_1 - \bar{X}_2}$ and t:

	Gender	
	Male (Category 1)	Female (Category 2)
	$\bar{X}_1 = 2.75$	$\bar{X}_2 = 3.01$
	$s_1 = 2.030$	$s_2 = 2.225$
	$N_1 = 855$	$N_2 = 1085$

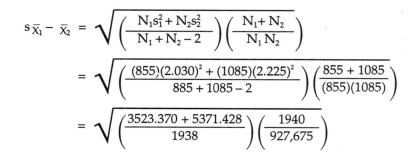

$$= \sqrt{(4.590)(.0021)}$$

$$= \sqrt{.0096}$$

$$= .098$$

and

$$t = \frac{(\overline{X}_1 - \overline{X}_2)}{s_{X_1} - \overline{X}_2}$$

$$= \frac{2.75 - 3.01}{.098}$$

$$= \frac{-.26}{.098}$$

$$= -2.653$$

Since we are interested only in a difference between males and females and not in the direction of any difference, we can ignore the negative sign and use $t = 2.653$. Appendix Table 2 presents values of t and associated probabilities. This table is set up much like the chi-square table we used in Chapter 6. Each row corresponds to the degrees of freedom listed in the leftmost column. Significance levels are in columns. Note that the table lists significance levels for what are called one-tailed tests and two-tailed tests. I'll explain the difference in Section 8.4. Meanwhile, just use the two-tailed levels, which is what we will do most of the time anyway. Each cell entry in the t table is the minimum value of t needed, for given degrees of freedom, to reject the null hypothesis at a given level of significance.

Here are the sections of the t table we need for this example—the two-tailed heading and very bottom rows:

	Level of Significance for Two-Tailed Test					
	.20	.10	.05	.02	.01	.001
.	.	.	.	.	.	.
.	.	.	.	.	.	.
.	.	.	.	.	.	.
120	1.289	1.658	1.980	2.358	☞ 2.617	3.373 ☜
∞	1.282	1.645	1.960	2.326	2.576	3.291

We have $855 + 1085 - 2 = 1938$ degrees of freedom, so we fall between the bottom row for infinite degrees of freedom and the row above it for 120 degrees of freedom. We'll be conservative (i.e., require the larger t value), using the 120 df row. The table tells us (▨) that we need a t of 3.373 or greater for statistical significance at the .001 level. Our actual t of 2.653 doesn't achieve significance at that level. But a t of 2.617 is needed for significance at the .01 level, and our t of 2.653 is slightly larger than that. Thus, with a significance level of .01, we reject the null hypothesis of no gender difference in daily television watching. We conclude that there is a relationship between gender and TV watching in the population from which our GSS data were drawn.

Keep in mind that the t test does not guarantee that our conclusions about the relationship between independent and dependent variables are correct. Just as with the chi-square test, we will be making a Type I error in rejecting the null hypothesis if the null hypothesis is "really" true, but the chances of such an error are less than one in a hundred (i.e., $p < .01$). What a significance test like the t test does is to give us a rational basis for our decision about a null hypothesis by describing the probability that we will commit a Type I error. As with the chi-square test, researchers conventionally use .05 as the cutoff for statistical significance. That is, we reject the null hypothesis only if the probability of our sample outcome is less than one in twenty. But also as noted with the chi-square test, there is nothing sacred about the .05 level of significance. Heavy reliance on the .05 level as a cutoff for statistical significance is strictly a matter of convention (albeit a widely honored one). Furthermore, researchers usually report even smaller probabilities such as $p < .02$, $p < .01$, or $p < .001$ if they are achieved.

8.3   Assumptions and Cautions Concerning t Test

I have noted a couple times that the t test assumes that the dependent variable is interval/ratio. After all, the t test assesses the difference between means, and calculation of means assumes a standard unit of measurement—that is, an interval/ratio variable. However, many researchers relax this interval/ratio requirement to also apply the t test to ordinal dependent variables. A researcher might, for example, use the t test to compare White and African-American attitudes measured on a strongly agree to strongly disagree scale. This relaxing of the level of measurement assumption is not surprising. I mentioned in Section 3.5 that means are sometimes informative with ordinal variables too.

But the t test is based on several assumptions in addition to an interval/ratio level of measurement. The t test assumes that the dependent variable is normally distributed in the population. Although a researcher typically does not know much about the population, fortunately this assumption can be relaxed for large samples. Most researchers feel comfortable applying the *t* test to samples of 50 or more even if the independent variable may not be normally distributed.

The *t* test also assumes that the two categories of the independent variable have equal dependent variable variances in the population (i.e., $\sigma_1^2 = \sigma_2^2$). This condition is called *homoscedasticity*. Fortunately, the *t* test is not affected much by moderate departures from homoscedasticity. If departures from homoscedasticity are more serious (usually indicated by very different sample variances), there are alternative techniques available that you will learn in more advanced statistics courses and texts.

We also assume (usually with good reason) that the sample means we are comparing are independent of one another. Independence is rarely a problem with most survey data like the General Social Survey since the selection of cases in one category of an independent variable is independent of the selection of cases in the other category. All cases in the GSS are randomly selected, and thus, in our example, the selection of males does not depend on the selection of females, and vice versa. There are many research designs, however, in which samples are not independent. Be especially careful of panel studies in which the same respondents are interviewed two or more times. Be careful too with experimental data for matched pairs of experimental and control group subjects. You will learn procedures for handling these more complex research situations in advanced statistics courses and texts.

A caution regarding *t* tests: With large numbers of cases, differences between means are likely to be statistically significant even if they are substantively trivial. Statistical significance does not imply substantive significance. Just as with the chi-square test, the outcome of a *t* test of the difference between means depends on both the magnitude of the difference and the number of cases. With enough cases, even a tiny, substantively trivial difference between means will be statistically significant. Our confidence in findings is rightfully greater if there are a large number of cases, but confidence doesn't imply that differences between means are worth paying much attention to. So we must be careful with *t* tests, both those of other researchers and our own. Don't let others bamboozle you about statistical significance when their Ns are large, and don't fool yourself either when your own Ns are large. Always assess substantive as well as statistical significance.

A few words on *t* tests and population data: Our consideration of *t* tests has concerned generalizing from differences between means in sample data to the population from which the sample was drawn. Many researchers, however, also apply difference of means tests to population data. They do so not to generalize from a sample, of course, but rather to rule out random processes as a likely source of whatever differences they find between independent variable categories in the populations being studied. (Once again, as with chi-square tests, your instructor may have a different perspective on the appropriateness of *t* tests applied to population data.)

8.4   One-Tailed and Two-Tailed Tests

We have so far been interested only in whether or not means were different, regardless of which mean was larger. There was no expectation or prediction about the direction of the difference between means. Our interest, for example, was in whether males and females differ in amount of TV watching. Of course, our comparison of sample means indicated which gender watches TV more, but we made no prior prediction about the direction of any male-female difference. Since our concern was only difference and not also the direction of the difference, we used both ends of the sampling distribution. That is, either a large negative *t* or a large positive *t* would lead us to reject the null hypothesis of no population difference. Statisticians and researchers call such tests *two-tailed tests* because they use both tails of the sampling distribution.

Two-tailed tests have become customary—the default, as it were. But research situations in which a researcher has expectations or predictions about the direction of the outcome may call for a *one-tailed test* that uses only one end of the sampling distribution. Suppose, for example, that a researcher hypothesizes that students who take part in study groups will do better on a statistics test than students who study alone. The researcher compares mean test scores of "group-study" students and "study-alone" students. Testing the statistical significance of the difference between means, the researcher is interested only in whether the group-study students score significantly higher than the study-alone students, and thus focuses only on the positive tail of the sampling distribution. The researcher certainly has no interest in group study producing lower mean test scores. It is not simply a difference in means that matters, therefore, but rather a difference that favors the group-study students.

To carry out a two-tailed test, we use the alpha value from the *t* table that describes twice the proportion as the analogous one-tailed test. For example, a two-tailed test at the .05 level requires the same *t* to reject the null hypothesis as a one-tailed test at the .025 level. There is no difference between one-tailed and two-tailed tests in calculations of *t* or degrees of freedom. This implies that if the direction of the difference between means is in the predicted direction, it is easier to reject a null hypothesis with a one-tailed test than with a two-tailed test. In this sense, two-tailed tests are more conservative than one-tailed tests.

Suppose, for example, that a researcher hypothesizes that students who study in a group will do better than students who study alone. On a statistics exam the researcher finds means of 84.5 for 16 group-study students and 81.3 for 16 study-alone students. That's a difference in the predicted direction, with group-study students scoring higher on the exam. So far so good for the hypothesis. Then the researcher calculates a *t* of 1.829 and 30 degrees of freedom. (We'll skip the actual calculations here.) The researcher then uses a one-tailed *t* test since direction is predicted. Going to the *t* table and using the one-tailed test columns, we see that with $N_1 + N_2 - 2 = 30$ degrees of freedom, a *t* of 1.697 is needed to reject the null hypothesis at the .05 level. Therefore, the difference between means is statistically significant at the .05 level, and the researcher rejects the null hypothesis of no difference between group-study and study-alone students. Note, however, that had a two-tailed test been used, a *t* of 2.042 or more would be needed to reject the null hypothesis. Thus, with a two-tailed test, the *t* of 1.697 would not have been statistically significant at the conventional .05 level and the researcher would not have rejected the null hypothesis of no difference between means.

As a general rule, however, researchers use two-tailed tests unless there is good reason for a one-tailed test, and you should do likewise.

8.5   Confidence Intervals for Differences Between Means

It may have occurred to you that testing for the significance of a difference between means has some connection to establishing confidence intervals around the mean, as we did in Section 4.6. Both statistical procedures involve means and both involve significance/confidence levels like .05 or .01. Moreover, for large Ns the *t* distribution used for the difference between means approximates the normal distribution used to establish a confidence interval around a mean. Yes, confidence intervals around means and the *t* test for differences between means have a lot in common.

We can easily establish *confidence intervals around a difference between means* by extending Section 4.6's formulas for confidence intervals around the mean to apply to differences between means:

$$95\% \text{ confidence interval} = (\overline{X}_1 - \overline{X}_2) \pm t_{.05} s_{\overline{x}_1 - \overline{x}_2}$$

$$99\% \text{ confidence interval} = (\overline{X}_1 - \overline{X}_2) \pm t_{.01} s_{\overline{x}_1 - \overline{x}_2}$$

More generally, the confidence interval around the difference between means is given by

$$(1 - \alpha) \text{ confidence interval} = (\overline{X}_1 - \overline{X}_2) \pm t_\alpha s_{\overline{x}_1 - \overline{x}_2}$$

where α = the alpha level corresponding to the confidence level

t_α = the value of t associated with the α level for given degrees of freedom $(N_1 + N_2 - 2)$

$s_{\overline{x}_1 - \overline{x}_2}$ = the standard error of the difference between means

Interpretation of a confidence interval for the difference between means is analogous to interpretation of the confidence interval around a mean. The chances are 95 out of 100 that the population difference between means lies within a 95 percent confidence interval. The chances are 99 out of 100 that the population difference between means lies within a 99 percent confidence interval.

Confidence intervals and analogous t tests for the differences between means thus give the same results with respect to statistical significance. So, for example, if a t test indicates significance at the .05 level, then a 95 percent confidence interval around the difference between means will not include 0. After all, if the t test indicates that less than 5 percent of possible samples will have a difference between means of 0 if there is no difference between means in the population, then the confidence interval should also indicate that 95 percent of samples will not include a difference between means of 0. Likewise, a 99 percent confidence interval around a difference between means that is significant at the .01 level will not include 0.

Take the gender → TV watching example of Section 8.1, where we found a male-female difference between means of −.26 and a standard error of .098. A t test, carried out in Section 8.2, found the male-female difference significant at the .01 level (t = 2.653, with df =1938). Here's the 99 percent confidence interval for the difference between means:

$$99\% \text{ confidence interval} = (\overline{X}_1 - \overline{X}_2) \pm t_{.01} s_{\overline{x}_1 - \overline{x}_2}$$

$$= (2.75 - 3.01) \pm (2.576)(.098)$$

$$= -.26 \pm .25$$

$$= -.51 \text{ to } -.01$$

Thus, the chances are 99 out of 100 that the difference between means in the population lies between −.51 and −.01. Note that 0 does not fall within that interval. However, the 99.9 percent confidence interval corresponding to a .001 significance level does include 0:

$$99.9\% \text{ confidence interval} = \left(\overline{X}_1 - \overline{X}_2\right) \pm t_{.001}s_{\overline{X}_1 - \overline{X}_2}$$

$$= (2.75 - 3.01) \pm (3.291)(.098)$$

$$= -.26 \pm .32$$

$$= -.58 \text{ to } .06$$

The chances are 99.9 out of 100 that the difference between means for males and females in the population is more than −.58 and less than .06. Note that 0 falls within that interval.

It is worth knowing that since the *t* distribution approaches the normal distribution as N increases, for large Ns the frequently used 95 and 99 percent confidence intervals become:

$$95\% \text{ confidence interval} = \left(\overline{X}_1 - \overline{X}_2\right) \pm 1.96s_{\overline{X}_1 - \overline{X}_2}$$

$$99\% \text{ confidence interval} = \left(\overline{X}_1 - \overline{X}_2\right) \pm 2.58s_{\overline{X}_1 - \overline{X}_2}$$

Here once again are those values—1.96 and 2.58—that we initially encountered in Chapter 4 when we first met normal curves and confidence intervals.

8.6   Summing Up Chapter 8

Here is what we have learned in this chapter:

- A *t* test is used to assess the statistical significance of differences between two means.

- The distribution of *t* approaches a normal distribution as sample sizes increase.

- The *t* test assumes independence of the sets of scores in comparison groups, homoscedasticity (equal variances of the comparison groups in the population), and normal population variances.

- This assumption of normal population variances can be relaxed when N is 50 or larger.

- Statistical significance is not the same as substantive significance, and statistical significance may be due to large Ns rather than substantively important differences between means.

- One-tailed tests can be used when the direction of the difference between means is predicted.

- Two-tailed tests are more conservative than one-tailed tests and are usually preferred.

• Confidence intervals can be established around differences between means. As tests of statistical significance, these intervals give the same results as analogous *t* tests.

Key Concepts and Procedures

Ideas and Terms

difference of means test
box-and-whisker diagram (boxplot)
t test
t (Student's *t*)
homoscedasticity

one-tailed tests
two-tailed tests
confidence interval around a
   difference between means

Symbols

$\mu_1 - \mu_2$

$\overline{X}_1 - \overline{X}_2$

t

$s_{\overline{X}_1 - \overline{X}_2}$

Formulas

$$t = \frac{(\overline{X}_1 - \overline{X}_2)}{s_{\overline{X}_1 - \overline{X}_2}}$$

$$s_{\overline{X}_1 - \overline{X}_2} = \sqrt{\left(\frac{N_1 s_1^2 + N_2 s_2^2}{N_1 + N_2 - 2}\right)\left(\frac{N_1 + N_2}{N_1 N_2}\right)}$$

$$df = N_1 + N_2 - 2$$

95% confidence interval $= (\overline{X}_1 - \overline{X}_2) \pm 1.96 s_{\overline{X}_1 - \overline{X}_2}$

99% confidence interval $= (\overline{X}_1 - \overline{X}_2) \pm 2.58 s_{\overline{X}_1 - \overline{X}_2}$

$(1 - \alpha)$ confidence interval $= (\overline{X}_1 - \overline{X}_2) \pm t_\alpha s_{\overline{X}_1 - \overline{X}_2}$

ANALYSIS WRITE-UP 4: COMPARISON OF MEANS AND *t* TEST

✦   ✦   ✦   ✦   ✦

Comparison of means and an associated t test for a single relationship can be embedded entirely within the text of a report or paper—no need for a table. However, tables are more effective for presenting a set of comparisons of means, especially for related dependent variables. The following write-up compares men and women on various feelings and emotions.

The General Social Survey asked respondents how many days in the past seven days they felt various sentiments—calm, outrage, happy, and the like. Table 1 reports means by gender along with a *t* test of the difference between means. Women report somewhat fewer days feeling calm and more days feeling sad, lonely, and worried. These differences are all statistically significant except for feelings of outrage, happiness, and shame.

TABLE 1 ABOUT HERE

Include the table at the end of the paper or report:

Table 1. Mean Days in Past Week with Various Feelings by Gender

Feelings	Gender		*t*	p
	Men	Women		
Calm	4.80	4.39	3.299	.001
	(632)	(817)		
Outraged	1.48	1.52	.371	n.s.
	(633)	(815)		
Happy	5.34	5.21	1.162	n.s.
	(636)	(815)		
Sad	1.42	1.81	3.910	.001
	(633)	(816)		
Ashamed	.47	.47	.057	n.s.
	(634)	(816)		
Lonely	1.28	1.67	3.362	.001
	(634)	(816)		
Worried	2.43	3.12	4.848	.001
	(634)	(816)		

CHAPTER 9
Analysis of Variance

In the last three chapters we learned to apply a *t* test to assess the statistical significance of the differences between means. This works fine with a dichotomous independent variable like gender. But what about categorized independent variables that have three, four, five, or even more values? In this chapter we'll turn to such bivariate situations in which we have a nondichotomous but categorized independent variable measured at any level and an interval/ratio dependent variable. We will learn a technique—analysis of variance—that takes full advantage of the interval/ratio level of the dependent variable.

For example, we might be interested in the effect of education on the amount of time people watch television. In Section 5.4 we collapsed the General Social Survey variable for hours per day of television watching into four categories and then analyzed its relationship to educational level using cross-tabulation techniques (see Table 5.8). As we have seen several times, however, collapsing interval/ratio scores into ordinal categories loses information about differences among individual scores. *Analysis of variance* is a statistical technique that makes full use of the interval/ratio level of dependent variables like television watching measured in hours per day. Analysis of variance is widely used enough to have an acronym: *ANOVA*. We will learn ANOVA in this chapter.

After this chapter you will be able to:

1. Interpret a box-and-whisker diagram for nondichotomous independent variables.

2. Understand the purpose of analysis of variance.

3. Understand and recognize conditions under which analysis of variance is appropriate or inappropriate.

4. Explain and calculate total, between-groups, and within-groups sums of squares.

5. Use a table of *F* values and interpret the statistical significance of *F* scores.

6. Construct and interpret an ANOVA table.

7. Interpret eta squared (E^2) as a proportional reduction in error (PRE) measure of association.

8. Recognize that ANOVA results depend in part on sample size and not confuse statistical significance with substantive significance.

9. Recognize, once again, that association does not imply causation.

9.1   Box-and-Whisker Diagrams/Differences Among Means

Table 9.1 lists the educational levels and amounts of daily TV watching of a sample of hypothetical respondents. To keep matters simple, I am using only 10 cases and only three education categories. A few paragraphs from now you'll be glad I'm keeping this example simple. I have spaced these 10 cases to call attention to the three categories of education—less than high school, high school, and college.

Table 9.1. Education and Hours Per Day Watching Television (10 Hypothetical Respondents)

Case Number	Education	TV Watching (Hours/Day)
01	Less Than H.S.	3
02	Less Than H.S.	4
03	High School	2
04	High School	2
05	High School	3
06	High School	4
07	High School	4
08	College	1
09	College	2
10	College	3

In Chapter 8 we learned to use box-and-whisker diagrams for dichotomous independent variables. Extension of box-and-whiskers to nondichotomous variables is straightforward. Figure 9.1 displays the

10-case Education → TV watching data as a box-and-whisker diagram. As in Chapter 8, vertical arrays of data points show the distributions of dependent variable (TV watching) scores within each category of the independent variable (Education) listed along the horizontal axis. Even for this very small data set, a few data points represent a couple cases superimposed on top of one another. The two high school grads who each watch TV 2 hours a day are superimposed. Likewise, the two high school grads who each watch 4 hours of TV a day.

Figure 9.1.   Box-and-Whisker Diagram for Daily Television Watching by Education

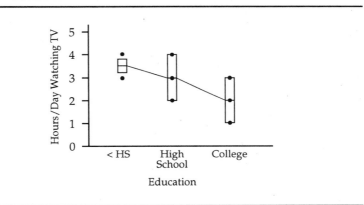

As with dichotomous independent variables, the elongated boxes here extend one standard deviation above and one standard deviation below mean dependent variable scores for values of the independent variable. We see in Figure 9.1, for example, that respondents with less than a high school education have a smaller standard deviation and thus a smaller variance than either high school or college graduates. Because of the small number of cases and particular scores in this example, boxes representing standard deviations in Figure 9.1 enclose all cases in two independent variable categories, but this is not generally the situation. We'll see more dispersed patterns with whiskers extending beyond boxes as we learn ANOVA and work with real data in the workbook. In our example, we are using so few cases that the whiskers are trimmed down to the boxes.

The mean TV watching score for each education category is represented by the horizontal line in the middle of each box. The line connecting the boxes joins the mean TV watching scores of education groups. If the independent variable is nominal, the ordering of groups is arbitrary and we cannot attach any meaning to the up and down direction of the line (except to note that means vary across groups). However, if the independent variable is interval/ratio or ordinal (as in this example), the line connecting box midpoints indicates the direction of the relationship between the two variables—positive, negative, curvilinear, or a more complex pattern. Here the line slopes down from the left to right, reflecting decreasing TV watching as we move from the lowest to highest educational levels. TV watching is highest among the less educated and lowest among the most educated—a negative relationship.

In principle, we could use cross-tabulation to examine the relationship between education and TV watching. Obviously in this example we would run into the practical problem of very tiny numbers of cases on which to base percentages, but with variables of this sort for a large data set like the General Social Survey, cross-tabulation is a possibility. Still, cross-tabulation would require collapsing the interval/ratio dependent variable—hours of TV watching—into a few categories. Instead of using cross-tabulation and thus losing this independent variable information by collapsing, we can compute the TV watching mean for each education category, and then compare means to determine if TV watching is related to education.

We have already seen means displayed in the box-and-whisker diagram, but let's actually compute them. The mean TV watching for our two cases with less than a high school education is 3.5 hours per day (i.e., $\frac{3+4}{2}$ = 3.5). The high school category's mean is 3.0 hours. And our college category watches TV an average of 2.0 hours each day. We might also note, in passing, that the total mean (i.e., the mean for all 10 cases) is 2.8 hours a day. Therefore, what we find are the TV watching means for each educational category presented in Table 9.2. I have also included standard deviations in Table 9.2. We'll find later on that including standard deviations is conventional in tables of this sort to describe the variation within independent variable categories.

Table 9.2. Mean Daily Television Watching
by Education

TV Watching	Less Than Hi School	High School	College	Total
	Education			
Mean Hours/Day	3.5	3.0	2.0	2.8
Standard Deviation	0.7	1.0	1.0	1.1

You don't have to be a three-time *Jeopardy!* champion to see that the two variables are negatively related. The more education, the less television watching. Of course, we have already seen this pattern displayed visually in the box-and-whisker diagram of Figure 9.1. But now we have a table of actual means to present and work with.

To generalize the example: We calculate dependent variable means within categories of the independent variable, and we then compare means across categories of the independent variable. Sound familiar? It certainly should. This procedure is directly analogous to cross-tabulation. For percentage tables we calculate *percentages* within and compare *percentages* across categories of the independent variable. Now we calculate *means* within and compare *means* across categories of the independent variable. The logic is exactly the same. In effect, means replace the percentages used in cross-tabulation. The logic is exactly the same too as the comparison of two means described in Chapter 8. The only difference is that here we have more means.

9.2   Purpose and Assumptions of Analysis of Variance

All this is very easy. What is not so easy is deciding whether or not the differences in TV watching across education categories are statistically significant. Can we generalize from a sample to the population? Did the differences we found happen just by chance, or can we conclude with some confidence that the differences in means we find in a sample exist in the population? What we need is a procedure to test for the differences among means analogous to the chi-square test for cross-tabulation or the *t* test for the difference between two means. Statisticians have figured out such a procedure: analysis of variance, or ANOVA.

Analysis of variance requires several assumptions:

1. *Interval/Ratio Dependent Variable:* We assume that the dependent variable is measured at the interval/ratio level. Hours of TV watching per day meets this assumption. Many social scientists, however, feel comfortable using ANOVA with ordinal dependent variables as well, just as they use means to describe variables in a univariate analysis.

2. *Random Sampling*: We must assume that the sample is drawn randomly from a population. Most well-conducted surveys like the General Social Survey meet this assumption reasonably well.

3. *Independence:* We must assume that the category means are independent of one another. Applied to our example, this assumption requires that the scores of the three education groups—less than high school, high school, and college—are sampled independently. This assumption is usually met by cross-sectional

survey data such as the General Social Survey.[1] After all, the sampling of any case or set of cases is independent of the sampling of any other cases. (For example, the chance of your Aunt Betty being selected for the GSS has nothing to do with the chance your stat instructor is selected.)

4. *Normal Distributions:* We must assume that the dependent variable is normally distributed in the population, although analysis of variance is robust enough to withstand some departure from normality, especially with large samples.

5. *Homoscedasticity:* We must assume that the distributions of the dependent variable within independent variable categories have equal variances in the population—a condition statisticians call **homoscedasticity**.

These five conditions underlie the model on which analysis of variance is based. To be sure, we rarely meet all five conditions completely and researchers sometimes necessarily bend these assumptions a little. As noted with the first assumption, many researchers use ANOVA with ordinal dependent variables. Violating assumptions underlying ANOVA, however, entails risks of misleading results.

9.3   The Logic of Analysis of Variance

We're going to take an intellectual trip now, so put on your thinking cap.

Analysis of variance works by breaking the total variation in dependent variable scores into two parts: the variation that occurs *within* each independent variable group and the variation that occurs *between* independent variable groups. For example, we break the total variation in TV-watching scores into the variation of scores within each of the three education groups and the variation between the education groups. If variables are associated, we expect independent variable groups to differ quite a bit from one another and, therefore, we expect the variation between groups to be greater than the variation within groups.

A little thought reveals that this approach generalizes the relationship of individual dependent variable scores to the total mean. I use *total mean* throughout this chapter to refer to the grand or overall

---

[1] Cross-sectional data are collected at a single point in time. The assumption of independent means is not met, however, by repeated measurements of the same cases at two or more times (e.g., panel data). A case's score at one time is not independent of the same case's scores at other times. See advanced statistics texts for ANOVA procedures for handling such situations.

mean of all scores—in our Education $\rightarrow$ TV watching example, 2.8 hours. The first of our 10 cases is a person who did not graduate from high school who watches TV 3 hours a day. This case deviates .2 hour from the total mean (i.e., $3 - 2.8 = .2$) and $-.5$ hour from the mean for the less than high school group ($3 - 3.5 = -.5$). The mean for the less than high school group itself deviates .7 from the total mean (i.e., $3.5 - 2.8 = .7$). Note that $.2 = .7 + (-.5)$. Equivalent patterns can be found for all cases.

In general, then, for each case:

Deviation of Dep. Var. Score from Total Mean	=	Deviation of Dep. Var. Score from Group Mean	+	Deviation of Group Mean from Total Mean

But that's a bit wordy, so let me use some shorthand notation:

$$(X_i - \overline{X}_T) = (X_i - \overline{X}_G) + (\overline{X}_G - \overline{X}_T)$$

where X_i = dependent variable score of the ith case

$\overline{X}_T$ = total mean (i.e., the mean of all cases)

$\overline{X}_G$ = mean of the Gth independent variable group that the ith case is in

Each of the parenthetical expressions in this equation has a clear interpretation. $(X_i - \overline{X}_T)$ measures the deviation of a score from the total mean. $(X_i - \overline{X}_G)$ measures the deviation of a score from its group mean (e.g., the mean for cases with less than a high school education). This is the distance between each data point and the midpoint of its box in Figure 9.1. And $(\overline{X}_G - \overline{X}_T)$ measures the deviation of the group mean from the total mean. You can easily see that the above equation is true if you remove the parentheses:

$$X_i - \overline{X}_T = X_i - \overline{X}_G + \overline{X}_G - \overline{X}_T$$

Then $-X_G$ and $+X_G$ cancel out and we are left with

$$X_i - \overline{X}_T = X_i - \overline{X}_T$$

Of course, we are interested in all scores rather than an individual score, so let's take all scores into account by measuring the variation for each of these three deviations. In Section 4.1 we learned to measure variation with a sum of squares—that is, a sum of squared deviations about the mean. Since each of the three expressions described above involves a deviation from a mean, let's square each one and add them across all cases. (We are interested here in *variation*, not the variance, so we are working only with sums of squares.)

Squaring and adding the $(X_i - \overline{X}_T)$ terms, we find $\Sigma(X_i - \overline{X}_T)^2$, the sum of squares that measures the variation of individual scores about the total mean. We call this the ***total sum of squares***. In our 10-case example:

$$
\begin{aligned}
\Sigma(X_i - \overline{X}_T)^2 &= (3 - 2.8)^2 + (4 - 2.8)^2 + (2 - 2.8)^2 + (2 - 2.8)^2 + (3 - 2.8)^2 \\
&\quad + (4 - 2.8)^2 + (4 - 2.8)^2 + (1 - 2.8)^2 + (2 - 2.8)^2 + (3 - 2.8)^2 \\
&= (.2)^2 + (1.2)^2 + (-.8)^2 + (-.8)^2 + (.2)^2 \\
&\quad + (1.2)^2 + (1.2)^2 + (-1.8)^2 + (-.8)^2 + (.2)^2 \\
&= .04 + 1.44 + .64 + .64 + .04 + 1.44 + 1.44 + 3.24 + .64 + 04 \\
&= 9.60
\end{aligned}
$$

The total sum of squares is 9.60 for the variation of scores about the total mean.

Likewise for $(X_i - \overline{X}_G)$, we have $\Sigma(X_i - \overline{X}_G)^2$, the sum of squares that measures the variation of individual scores about their group means. This sum of squares, in other words, measures the less-than-high-school TV watching scores' variation about their group mean, the high school graduate scores' variation about the high school mean, and the college scores' variation about the college mean. $\Sigma(X_i - \overline{X}_G)^2$ is called the ***within-groups sum of squares***. Here's our example:

$$
\begin{aligned}
\Sigma(X_i - \overline{X}_G)^2 &= (3 - 3.5)^2 + (4 - 3.5)^2 + (2 - 3)^2 + (2 - 3)^2 + (3 - 3)^2 \\
&\quad + (4 - 3)^2 + (4 - 3)^2 + (1 - 2)^2 + (2 - 2)^2 + (3 - 2)^2 \\
&= (-.5)^2 + (.5)^2 + (-1)^2 + (-1)^2 + (0)^2 + (1)^2 + (1)^2 \\
&\quad + (-1)^2 + (0)^2 + (1)^2 \\
&= .25 + .25 + 1 + 1 + 0 + 1 + 1 + 1 + 0 + 1 \\
&= 6.50
\end{aligned}
$$

The within-groups sum of squares is 6.50. It measures the deviations of cases' TV-watching scores about the means for Education categories.

Finally, expanding $(\overline{X}_G - \overline{X}_T)$ to measure the variation of group means about the total mean, we have $\Sigma N_G(\overline{X}_G - \overline{X}_T)^2$, where N_G is the number of cases in the Gth group. With its inclusion of the term N_G, this expression is a little tricky, so let's be careful. We *cannot* just subtract the total mean from each group mean, square the difference, and add these squared differences across groups. That procedure would fail to take into account the number of cases in each group. Instead, we weight each of the squared differences by the size of each group in order to give larger groups the greater importance that their size merits. We do this by multiplying the squared difference for each group by N_G, the number of cases in the group. Thus, the expression tells us to square the difference between each group mean and the total mean, then multiply that squared difference by the number of cases in that group, and, finally, add these products across all groups. The result is called the *between-groups sum of squares*. Here we go with our example:

$$\Sigma N_G(\overline{X}_G - \overline{X}_T)^2 = 2(3.5 - 2.8)^2 + 5(3.0 - 2.8)^2 + 3(2.0 - 2.8)^2$$
$$= 2(.7)^2 + 5(.2)^2 + 3(-.8)^2$$
$$= .98 + .20 + 1.92$$
$$= 3.10$$

The between-groups sum of squares measuring the variation of group means about the total mean is 3.10.

Note that the total sum of squares equals the within-groups sum of squares plus the between-groups sum of squares:

9.60 = 6.50 + 3.10

Or more generally:

$$\Sigma(X_i - \overline{X}_T)^2 = \Sigma(X_i - \overline{X}_G)^2 + \Sigma N_G(\overline{X}_G - \overline{X}_T)^2$$

That is:

$$\begin{matrix} \text{Total Sum} \\ \text{of Squares} \end{matrix} = \begin{matrix} \text{Within-Groups} \\ \text{Sum of Squares} \end{matrix} + \begin{matrix} \text{Between-Groups} \\ \text{Sum of Squares} \end{matrix}$$

This equation is interesting and important. Just as we can break the difference between an individual score and the total mean into the differences between the score and its group mean and between the group mean and the total mean, so too can we break the total variation into the within-groups variation plus the between-groups variation.

But consider this: Sums of squares depend not only on the amount of variation but also on the number of cases on which they are based. The more cases, the larger the sums of squares are likely to be because more differences are being added. We learned that in Section 4.1. That's why we calculate a variance by dividing a sum of squares by $N - 1$, thereby finding the unbiased average sum of squares.

So we will now convert these total, between-groups, and within-groups sums of squares into variances. It might seem that we should divide the sums of squares by $N - 1$, just as we calculate a variance with sample data. That's in fact what we do for the total sum of squares but not for the between-groups or within-groups sums of squares. Statisticians have proven that we can get good estimates of variances if we divide each sum of squares by these degrees of freedom:[2]

$$\text{Total df} = N - 1$$

$$\text{Within-Groups df} = N - k$$

$$\text{Between-Groups df} = k - 1$$

$$\text{where } N = \text{total number of cases}$$

$$k = \text{number of categories of the independent variable}$$

In our example, the degrees of freedom are 9 for the total sum of squares, 7 for the within-groups sum of squares, and 2 for the between-groups sum of squares. I'll let you check to be sure I calculated these numbers correctly.

Dividing each sum of squares by its degrees of freedom, we arrive at estimates of the three variances. We get:

$$\text{Total Variance} = \frac{\Sigma(X_i - \overline{X}_T)^2}{N - 1}$$

$$= \frac{9.60}{10 - 1}$$

$$= \frac{9.60}{9}$$

$$= 1.067$$

This equation is an old friend. It's the ordinary variance for sample data that we met in Section 4.1.

$$\text{Within-Groups Variance} = \frac{\Sigma(X_i - \overline{X}_G)^2}{N - k}$$

$$= \frac{6.50}{10 - 3}$$

$$= \frac{6.50}{7}$$

$$= .928$$

---

[2] Dividing sums of squares by these degrees of freedom provides unbiased estimates of population parameters. That is, the mean of each statistic for all possible samples is equal to the population parameter. The notion of bias assumes a familiarity with the logic of sampling and is a little beyond this text.

$$\text{Between-Groups Variance} = \frac{\Sigma N_G(\overline{X}_G - \overline{X}_T)^2}{k-1}$$

$$= \frac{3.10}{3-1}$$

$$= \frac{3.10}{2}$$

$$= 1.550$$

These estimated variances often are called *mean sum of squares* (or just *mean squares*), but that is a bit of a misnomer. They are not really averages (i.e., means of squares) since we divide the sums of squares by degrees of freedom rather than by N, but they are good estimates of the three variances.

Now let's think about what these three variances have to do with the relationship between the independent and dependent variables. In fact, let's think very hard about this since it is both important and somewhat complicated.

We are interested in whether education affects TV watching. If education has a big effect on TV watching, then the TV-watching means for the education groups should be very different from one another. One education group might have a very high TV-watching mean, another education group might be very low, the third somewhere in between. Hmmm . . . then there would be a lot of variation among group means, and so the between-groups variance would be very large. So, a strong relationship will produce a large between-groups variance. Most of the variation among scores is taking place between education groups rather than within them.

But what if the effect of education on TV watching is moderate? In that situation, we would find some moderate but not large differences among education group means. Maybe one group mean would only be somewhat high, another only a little low, the third somewhere in the middle. There would be some variation among means, but not a great deal, and thus the between-groups variance would be moderate. Thus, a moderate relationship produces a moderate between-groups variance, indicating that a moderate amount of the variation in TV-watching scores is occurring between rather than within education groups.

And what if education has little or no effect on TV watching? Then the TV-watching means of education groups would be very similar. There wouldn't be much variation among them. And so for a very weak relationship, the between-groups variance would be very small. Therefore, most of the variation in TV watching is occurring within rather than between education groups.

I have summarized the ratio of between-groups variance to within-groups variance in Table 9.3. If you are still thinking very hard, you see that the pattern of ratios in Table 9.3 suggests that we can assess the overall strength of a relationship by forming a ratio of between-groups variance to within-groups variance, as shown in the right-hand column. The stronger a relationship, the larger the ratio. The weaker a relationship, the smaller the ratio. This very important ratio is called the **F ratio,** or simply **F**.[3] Stated formally:

$$F = \frac{\text{Between-Groups Variance}}{\text{Within-Groups Variance}}$$

For our Education–TV-watching relationship, the F ratio is $\dfrac{1.550}{.928} = 1.67$.

Table 9.3. Ratios of Between-Groups Variance to Within-Groups Variance

Strength of Relationship	Between-Groups Variance	Within-Groups Variance	$\dfrac{\text{B-G Variance}}{\text{W-G Variance}}$	
Strong	Large	Small	$\dfrac{\text{Large}}{\text{Small}}$	= Big Ratio
Moderate	Medium	Medium	$\dfrac{\text{Medium}}{\text{Medium}}$	= Medium-Size Ratio
Weak	Small	Large	$\dfrac{\text{Small}}{\text{Large}}$	= Little Ratio

Now the conclusion: Statisticians have figured out the sampling distribution of F and, therefore, they know the probability of finding a given F. In other words, statisticians know the magnitudes of F needed to establish statistical significance at various levels. Table 3 in the Appendix presents the minimum F ratios necessary for significance at the .05, .01, and .001 levels. These F ratios and associated probabilities are based on the assumptions underlying ANOVA that I mentioned earlier: random sampling, independent means, an interval/ratio dependent variable that is normally distributed in the population, and homoscedasticity (i.e., equal variances).

The probability associated with an F ratio depends on the degrees of freedom. We have already calculated degrees of freedom a few pages back:

---

[3] This **F** has nothing to do, of course, with the F introduced in Chapter 2 for cumulative frequencies.

$$\text{Within-Groups df} = N - k$$
$$= 10 - 3$$
$$= 7$$
$$\text{and Between-Groups df} = k - 1$$
$$= 3 - 1$$
$$= 2$$

Because the significance of F ratios depends on two kinds of degrees of freedom, the F table is larger than tables for chi square or t values. In fact, it takes a separate page for each significance level. Appendix Table 3 of F values is divided into separate parts for the commonly used .05, .01, and .001 levels of significance. Degrees of freedom for the between-groups variance are listed along the top of the F table; degrees of freedom for the within-groups variance are listed down the left-hand column. The cell corresponding to the degrees of freedom tells us the minimum F value needed for significance at a given level. An F larger than the F in the cell means that the relationship is statistically significant at least at the given level. Here are the heading and the 7 df row from Appendix Table 3 for the .05 level:

N_1	N_2									
	1	2	3	4	5	6	8	12	24	∞
.	.	.	.	.	.	.	.	.	.	.
.	.	.	.	.	.	.	.	.	.	.
7	5.59	☞ 4.74	4.35	4.12	3.97	3.87	3.73	3.57	3.41	3.23

The hand points to the entry corresponding to 2 and 7 degrees of freedom. We need an F of 4.74 or more for significance at the .05 level. With an actual $F(2, 7)$ of 1.67, the relationship between daily hours of TV watching and education is *not* statistically significant, even at the conventional .05 level. Had the F ratio been significant at the .05 level, however, we would then have checked for significance at the .01 level. If significant at that level, we would then have checked for significance at the even more stringent .001 level. What we seek, in short, is the highest level of significance (i.e., smallest probability) that our F ratio achieves.

But in this case, our assessment of statistical significance ends at our first stop—the .05 F table. Our $F(2, 7)$ of 1.67 is less than the 4.74 required for the .05 level and, therefore, is not statistically significant. The F test tells us that education and TV watching are not really related in the larger population (at least in this hypothetical example).

There is a pretty high probability that our sample's differences among the three education categories' TV watching means could have occurred just by chance even if there are no differences among category means in the population. And that's what we wanted to find out when we began this section. We have tested for the statistical significance of the differences among means and found that they are not significantly different. Stated differently but equivalently: Education does not explain a statistically significant proportion of the variation in TV watching. And stated still another but equivalent way: We cannot reject the null hypothesis of no relationship between education and TV watching. Sure, we may make an error—a Type II error—in not rejecting the null hypothesis if it is false. But if we rejected the null hypothesis, we would risk making an error—a Type I error—more than one out of twenty times, and that is too high a risk for us to accept in most social scientific analyses.

By the way, you may have noticed two paragraphs back that I expressed the F in our analysis with the degrees of freedom following in parentheses, like this: $F(2, 7)$. It is conventional when reporting an F to include the between-groups and within-groups degrees of freedom this way.

I hope you realize that the logic of testing the statistical significance of differences among means is exactly the same as the logic underlying chi-square tests for bivariate tables or t tests for differences between means. F in an ANOVA is analogous to χ^2 or t. Like the distribution of χ^2 or t, the distribution of F is a sampling distribution. For given degrees of freedom, the distribution of F tells us the proportion of samples that would have differences among means as large as the differences we find in our sample if there were "really" no differences among means in the population. When that proportion is less than 1-in-20 (i.e., $p < .05$), we reject the null hypothesis of no differences among means in the population.

You may be sensing that ANOVA's F may have an even closer connection with the t statistic of Chapter 8. Indeed. Just as the statistic t tests for the statistical significance of differences between two means, so the F statistic tests for the significance of differences among three or more means. But the connection between F and t is even more intimate. For dichotomous independent variables (and ignoring the sign if t is negative), $t = \sqrt{F}$, with $N - 2$ degrees of freedom. Therefore, t and $\sqrt{F}$ are interchangeable.

So, with a dichotomous independent variable, if you compute F and take its square root, you will have the same number that you get computing t (disregarding any negative sign). With dichotomous independent variables, however, researchers usually report their results as a difference of means test, with a t test to assess statistical significance. For independent variables with three or more values, researchers report ANOVAs.

9.4   The ANOVA Table

Before presenting the results of an ANOVA, researchers usually present and discuss a table of means like Table 9.2. Such a table often includes standard deviations of the dependent variable within independent variable categories. Then the ANOVA itself is summarized in an *ANOVA table*. I will show you one so that you will know how to present your own analyses of variance. Table 9.4 presents an ANOVA table for our Education → TV watching example. Note that the probability is reported as n.s. for not significant. Had conventional significance levels been achieved, the table would read $p < .05$, $p < .01$, $p < .001$, or whatever the level of significance. There are several "standard" formats for ANOVA tables, so you may find tables that look a little different from this one. It is conventional, however, to omit the total mean sum of squares since it does not enter into the calculation of the F ratio.

Table 9.4. Analysis of Variance of Daily Hours of Television Watching by Education

Source	Sum of Squares	df	Mean Sum of Squares	F	p
Between Groups	3.10	2	1.550	1.67	n.s.
Within Groups	6.50	7	.928		
Total	9.60	9			

9.5   The Correlation Ratio (E^2)

ANOVA is certainly helpful. But let's go even further since there is additional information easily obtained from the sums of squares we have calculated. Let's think about the sums of squares. The total sum of squares is the total variation in our dependent variable, TV watching. It is the amount of variation of individual scores around the total TV-watching mean. If we were to guess how much each case watches television without knowing any education score, the best we could do is to guess the total mean. We would make errors, of course, and the total sum of squares is a measure of those errors.

But what if we know the education of each case? Then we could improve our guess of the TV watching of each case by guessing the mean TV watching for the case's education category. For example, for someone who is a high school graduate, we would guess that the person watches TV the mean number of hours watched by all high school graduates. We might actually increase our errors for some cases, but if education and television watching are related, then on average our guesses would reduce errors. Of course, there would still be errors—errors measured by the within-groups sum of squares. After all, that sum of squares is the variation going on within each education category. It is variation still left after we use information about education scores to estimate amounts of TV watching.

We have reduced our errors in guessing television watching by using our knowledge of education scores. But how much have we reduced errors? Answer: By the difference between the total sum of squares and the within-groups sum of squares.

However:

$$\text{Total Sum of Squares} = \text{Within-Groups Sum of Squares} + \text{Between-Groups Sum of Squares}$$

And so:

$$\text{Between-Groups Sum of Squares} = \text{Total Sum of Squares} - \text{Within-Groups Sum of Squares}$$

Therefore, the between-groups sum of squares is a measure of how much we reduce errors in guessing dependent variable scores if we know independent variable scores. Dividing the between-groups sum of squares by the total sum of squares expresses this error reduction as a proportion. This proportion is the *correlation ratio*, usually called *eta squared* and symbolized E^2 or η^2:

$$E^2 = \frac{\text{Between-Groups Sum of Squares}}{\text{Total Sum of Squares}}$$

In our example:

$$E^2 = \frac{\text{Between-Groups Sum of Squares}}{\text{Total Sum of Squares}}$$

$$= \frac{3.10}{9.60}$$

$$= .323$$

$$= .32$$

We reduce errors in estimating daily television watching 32 percent by using information about cases' levels of education—more specifically, by estimating TV-watching scores with the mean hours of TV watching for each level of education. (Remember, however, that this 32 percent is based on fictitious data.)

Eta squared is a measure of association. It describes how strongly the dependent variable is related to the independent variable. Indeed, E^2 is a proportional reduction in error (PRE) measure of association since it describes the proportion by which errors in guessing the dependent variable scores are reduced by knowledge of the independent variable. E^2 is like lambda and gamma in this respect although, of course, it is based on a different "guessing rule." E^2 gauges error reduction if we guess the mean dependent variable score for each category of the independent variable rather than guessing the total mean.

Another but equivalent way of expressing the same thing is to say that E^2 is the proportion of variation in the dependent variable explained by the independent variable. That is how researchers usually express E^2. With $E^2 = .32$, we say that education explains about 32 percent of the variation in hours per day of television watching.

9.6   Two-Way Analysis of Variance (and Beyond)

We have concerned ourselves with an analysis of variance involving a single independent variable. This procedure is called *one-way analysis of variance*. Analysis of variance can be generalized, however, to include additional independent variables. We might be interested, for example, in the effects of both education and gender on TV watching. Are there differences in average television watching not only among education groups, but also between males and females?

This generalization of ANOVA allows assessments of the effects of each independent variable on the dependent variable as well as the interaction effects of combinations of values of independent variables. Thus, we can use analysis of variance to assess, for example, the effects of education, the effects of gender, and the interactive effects of education and gender together. The latter might reveal, for example, that the combination of being both highly educated and a female has an effect on TV watching that could not be predicted by education and gender considered separately.

Interaction effects are common. Another example: Both alcohol and certain medications may impair activities, like driving, that require coordination and decision-making. But the effects of alcohol and medication taken together may be far greater than even the combined independent effects of alcohol and medication. The two independent variables exaggerate the effects of each other into a sort of double-whammy.

Techniques incorporating additional independent variables change names appropriately. For example, *two-way analysis of variance* examines the independent and interaction effects of—you guessed it—two independent variables. But the underlying logic is the same as for one-way analysis of variance. Once additional variables are introduced, of course, extensions of analysis of variance leave bivariate analysis and cross over into multivariate analysis. Two-way analysis of variance and beyond are a little outside the scope of this text, but I want you to know that such procedures are available. Also beyond what we can do in this text are MANOVA (multivariate ANOVA) techniques that handle two or more dependent variables that are related to one another. I hope you will learn these more sophisticated techniques in a more advanced text or course.

9.7   Three Cautions About Statistically Significant F Ratios

ANOVA tests for the statistical significance of differences among means. Finding a statistically significant F, however, does not necessarily mean that *all* the means are different from one another. ANOVA tests for an overall difference among means, not for differences between particular means. Statistical significance may be due to only one or two means that are very different from an otherwise similar set of means. Fortunately, statisticians have developed procedures called post-hoc tests to handle such situations. These tests identify the particular sources of statistical significance. Unfortunately, these procedures must wait until you get into more advanced statistical techniques than this text covers. Meanwhile, don't be timid but do be properly cautious when interpreting statistical significance in ANOVA. Be sure to compare actual means rather than relying on just F test results to tell you if means are different from one another.

As with all tests of statistical significance, the F test of differences among means depends on the number of cases. As always, statistical significance is easier to achieve with larger samples. This is quite proper—we certainly should have more confidence generalizing from large samples than small samples. But the proper dependence of F tests on sample size implies, once again, that we should not be overly impressed by results of F tests based on large samples. Statistical significance does not imply substantive significance. Be mindful of the effects of sample size on significance test results, and be sure to actually inspect means to see whether differences among them are substantively important.

Just as with cross-tabulation and difference of means, the finding of an association using ANOVA does not necessarily imply that the two variables are *causally* related. Why not? Because of exactly the same reasons suggested for cross-tabulations in Section 5.8. Recall the examples of storks and babies, Ganges flooding and street crime, clergy salaries and liquor consumption. The association described by ANOVA may be due to some other variable that affects both the variables in our bivariate analysis. Many pairs of variables are associated that are not *causally* related. We will take this matter up in some detail in Chapter 11.

9.8   Summing Up Chapter 9

Here is what we have learned in this chapter:

- Analysis of variance assesses the relationship between a categorized independent variable and an interval/ratio dependent variable.

- The total sum of squares measures variation of dependent variable scores around the total mean.

- The between-groups sum of squares measures variation of group means around the total mean.

- The within-groups sum of squares measures variation of scores around their category means.

- Analysis of variance examines differences among means by decomposing the total variance in the dependent variable into variance that occurs within independent variable groups and variance that occurs between independent variable groups.

- Analysis of variance assumes random sampling, independent means, an interval/ratio dependent variable normally distributed in the population, and equal population variances of the dependent variable within independent variable categories (i.e., homoscedasticity).

- The F ratio is the ratio of between-groups variance to within-groups variance.

- When an independent variable is dichotomous, an analysis of variance is equivalent to a t test of the difference between means, with $t = \sqrt{F}$.

- Eta squared (E^2) is a PRE measure of association that gauges the strength of the relationship between two variables in an analysis of variance. Eta squared is the proportion of variation in the dependent variable explained by the independent variable.

- Analysis of variance can be extended to additional independent variables.

- Two-way and higher levels of analysis of variance examine not only the effects of each independent variable, but also the interaction effects of combinations of independent variables.

- Results of F tests of differences among means are affected by numbers of cases and may be influenced by one or two extreme means.

- Two variables shown by an analysis of variance to be associated are not necessarily *causally* related.

Key Concepts and Procedures

Ideas and Terms

analysis of variance (ANOVA)
box-and-whisker diagram (boxplot)
homoscedasticity
total sum of squares
within-groups sum of squares
between-groups sum of squares
total variance
within-groups variance

between-groups variance
mean sum of squares (mean squares)
F ratio
ANOVA table
eta squared (correlation ratio)
one-way analysis of variance
two-way analysis of variance

Symbols

X_i
$\overline{X}_G$
$\overline{X}_T$
N_G
k
F
E and E^2

Formulas

$$\text{Total Variance} = \frac{\Sigma(X_i - \overline{X}_T)^2}{N - 1}$$

$$\text{Total df} = N - 1$$

$$\text{Within-Groups Variance} = \frac{\Sigma(X_i - \overline{X}_G)^2}{N - k}$$

$$\text{Within-Groups df} = N - k$$

$$\text{Between-Groups Variance} = \frac{\Sigma N_G(\overline{X}_G - \overline{X}_T)^2}{k - 1}$$

$$\text{Between-Groups df} = k - 1$$

$$F = \frac{\text{Between-Groups Variance}}{\text{Within-Groups Variance}}$$

$$E^2 = \frac{\text{Between-Groups Sum of Squares}}{\text{Total Sum of Squares}}$$

$$t = \sqrt{F}$$

Analysis Write-up 5: Analysis of Variance

✦   ✦   ✦   ✦   ✦

Here's a write-up that first compares means and then carries out an ANOVA describing the relationship between respondent's age at which first child was born and region in which respondent lived at age 16.

The General Social Survey asked respondents who have had children how old they were when their first child was born. One outlier (65 years old when first child was born) was excluded from the analysis. Age when first child was born has a mean of 23.5 and a standard deviation of 5.22. Region in which respondents lived at age 16 was collapsed into five categories—foreign-born plus the American Northeast, Midwest, West, and South.

Table 1 presents the region-specific mean ages at which respondents had their first child. Respondents from the South were somewhat younger and respondents who came of age abroad were somewhat older when they had their first child. Overall, however, differences between regions are small. An ANOVA, reported in Table 2, indicates that the differences among means yield an $F(4, 2036) = 8.378$, which is statistically significant at the .001 level. Eta squared, however, is only .02. There is, then, a statistically significant but substantively small difference among regions with regard to mean ages at which respondents had their first child.

TABLES 1 AND 2 ABOUT HERE

Include the tables at the end of the paper or report:

Table 1. Mean Respondent Age When First Child Was Born by Region at Age 16

Age When First Child Was Born	Where Respondent Was Living at Age 16					Total
	Northeast	Midwest	West	South	Foreign	
Mean Age	24.1	23.8	23.5	22.7	24.9	23.5
Standard Deviation	5.18	5.28	5.19	5.07	5.26	5.22
(N)	(425)	(561)	(286)	(659)	(110)	(2041)

Table 2 . ANOVA of Mean Respondent Age When First Child Was Born by Region at Age 16

Source	Sum of Squares	df	Mean Sum of Squares	F	p
Between Groups	899.3	4	224.82	8.378	.001
Within Groups	54637.6	2036	26.84		
Total	55536.9	2040			

CHAPTER 10
Regression and Correlation

In this chapter we learn to analyze relationships between two interval/ratio variables. We might be interested, for example, in the relationship between urbanism and fertility rates among the largest nations, or between years of education and annual income in the General Social Survey. We could, of course, categorize the independent variable and then compare means and carry out an analysis of variance as described in the previous chapter. If we dichotomize the independent variable, we could even carry out a difference of means test. Or we could categorize both variables and use the cross-tabulation techniques described in Chapters 5 through 7.

But I have pointed out several times that categorization always involves some loss of information as more detailed data are collapsed into broader categories. Moreover, tabular analysis requires fairly large numbers of cases to provide a stable base for percentages. Fortunately, there are techniques that avoid these problems by taking advantage of the information provided by the standard units of measurement used for interval/ratio level variables. In this chapter we take up two of these techniques—regression and correlation.

After this chapter you will be able to:

1. Create and interpret scatterplots.

2. Recognize linear and nonlinear relationships in scatterplots.

3. Recognize the limitations of scatterplots.

4. Find a regression line and regression equation and explain what they are.

5. Calculate and interpret the slope and intercept of a regression line.

6. Calculate and explain a correlation coefficient.

7. Interpret r^2 as proportion of variation explained.

8. Carry out and interpret tests of statistical significance for correlation coefficients.

9. Explain what residuals are.

10. Distinguish between and explain listwise and pairwise deletion of missing data.

10.1   Scatterplots

In Section 2.10 we mapped fertility rates (defined as average number of children per woman) for the 50 most populous countries. We saw high fertility rates in Africa and parts of Asia, and generally low fertility in Europe and East Asia. We also saw roughly similar patterns (although in the opposite direction) for the percentages of populations living in urban areas. That is, countries with greater urbanism seemed to have lower fertility rates. Maybe urbanism affects fertility rates. Table 10.1 presents the specific fertility rates and urban percentages for all 50 countries. Fertility rate is operationally defined as the average (mean) number of children born per woman.

Table 10.1. Percent Urban and Fertility Rates (50 Most Populous Countries)

Country	Fertility Rate	Percent Urban	Country	Fertility Rate	Percent Urban
Algeria	3.96	50	Nigeria	6.43	16
Argentina	2.72	86	North Korea	2.40	60
Australia	1.83	85	Pakistan	6.50	28
Bangladesh	4.55	14	Peru	3.22	71
Brazil	2.49	76	Philippines	3.45	44
Canada	1.84	77	Poland	1.97	62
China	1.85	28	Romania	1.83	54
Colombia	2.54	68	Russia	1.83	73
Egypt	4.35	45	Saudi Arabia	6.70	79
Ethiopia	6.88	15	South Africa	4.40	57
France	1.80	74	South Korea	1.64	74
Germany	1.40	85	Spain	1.38	78
Great Britain	1.83	92	Sri Lanka	2.13	22
India	3.57	26	Sudan	6.19	23
Indonesia	2.86	31	Taiwan	1.81	75
Iran	6.40	57	Tanzania	6.25	21
Iraq	6.86	70	Thailand	2.16	19
Italy	1.37	68	Turkey	3.30	61
Japan	1.54	77	Uganda	7.15	11
Kenya	6.06	25	Ukraine	1.82	68
Malaysia	3.54	51	United States	2.05	75
Mexico	3.25	71	Uzbekistan	3.78	40
Morocco	3.96	47	Venezuela	3.14	84
Myanmar (Burma)	3.70	25	Vietnam	3.45	21
Nepal	5.33	8	Zaire	6.70	40

It looks from the maps in Section 2.10 as if these two variables are related. But appearances can deceive (ever see *Tootsie* or *The Crying Game*?), and certainly we need a more precise assessment of just how similar maps are than our subjective interpretation offers. We need a good, objective way to more explicitly describe the relationship between fertility and urbanism. Is there *really* a relationship between these two variables? Do more urban countries actually have lower fertility rates? And if so, how strong is the relationship? In posing these questions, I am asking about *countries*, not individuals, so I am *not* asking if urban people have fewer children than their rural brothers and sisters. That is a different question to be answered with data for individuals. We will stick to the ecological level here (and avoid the ecological fallacy described in Section 1.9).

Scatterplots (sometimes called *scattergrams*) are graphs that help us visualize relationships between interval/ratio variables. To construct a simple scatterplot, we draw a graph with intersecting perpendicular lines that displays the independent variable on the horizontal axis and the dependent variable on the vertical axis. Figure 10.1 shows the form of a scatterplot with percent urban as the independent variable and fertility rate as the dependent variable before data points are entered. By convention, the independent variable is arrayed along the horizontal or *X-axis* and the dependent variable is arrayed along the vertical or *Y-axis*. Scatterplots are often (but not always) set up so that the X- and Y-axes cross at the *origin*, or the zero point for each variable.

Figure 10.1. Fertility Rate by Percent Urban (empty scatterplot)

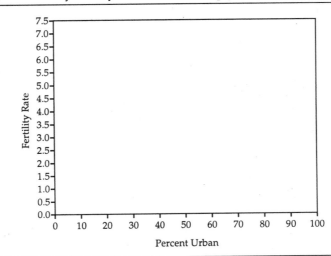

An empty scatterplot begs to be filled in. We do so by putting a symbol (a dot will do nicely) at each point on the graph corresponding to a case's scores on the independent and dependent variables. For example, Egypt, which is 45 percent urban and has a fertility rate of 4.35, receives a dot located 45 units to the right and 4.35 units up from the origin, as shown in Figure 10.2. We have done this sort of plotting of data points before in box-and-whisker diagrams (see Section 9.1). A box-and-whisker diagram is, in fact, a special type of scatterplot.

Figure 10.2. Fertility Rate by Percent Urban (Egypt only)

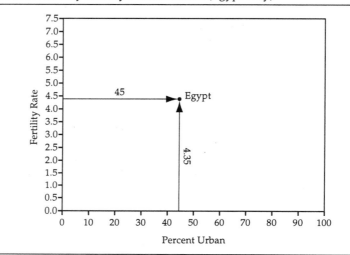

Continue adding dots until all 50 countries are located on the scatterplot, and we have the scatterplot shown in Figure 10.3.

Figure 10.3. Fertility Rate by Percent Urban (50 Most Populous Nations)

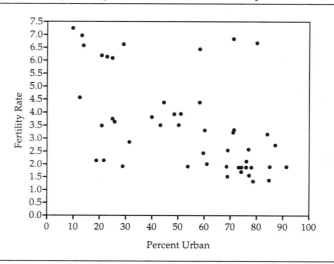

Scatterplots give us a good visual sense of relationships between interval/ratio variables. Often we can immediately identify the direction of a relationship. If a relationship is positive, scatterplot points form a general pattern from the lower left to the upper right. Lower dependent variable scores are associated with lower scores on the independent variable; and higher dependent variable scores are associated with higher scores on the independent variable. We'll see examples of positive relationships later.

If a relationship is negative, the general pattern of points sweeps from the upper left to the lower right. That pattern, after all, describes a negative relationship. Why? Because in the upper left low scores on the independent variable are associated with high scores on the dependent variable, and in the lower right high scores on the independent variable are associated with low scores on the dependent variable.

The general pattern of the relationship in Figure 10.3 between percent urban and fertility rate is negative. As urbanism increases, fertility decreases, and vice versa. Countries with lower urbanism generally have higher fertility rates. Countries with higher urbanism generally have lower fertility rates. We find only a few data points in the lower left or in the upper right of the scatterplot—exceptions to the general pattern. Most data points are somewhere around an imaginary diagonal from the upper left to the lower right. (Remember, however, that we are working with ecological data, so we cannot infer that urbanism and fertility are related at the individual level. Maybe they are, but we would need to analyze data from individual people in these countries to avoid committing an ecological fallacy.)

10.2   Scatterplots and the Strength of Relationships

Figure 10.4 presents generalized examples of scatterplots of positive and negative relationships. Think back to the patterns of positive and negative relationships in bivariate tables. Remember those schematic tables with Xs in the diagonals in Section 5.4? (If you don't, look back at them now, in Figure 5.1.) Those tables have the same general form as these positive and negative scatterplots. That is why I suggested that you set up tables with values of the column (independent) variable running from left to right and values of the row (dependent) variable running from high at the top to low at the bottom. That format corresponds to the way we arrange values on the axes of a scatterplot. In fact, although we won't bother to do so here, you could form a frequency table by drawing horizontal and vertical lines through a scatterplot to categorize the variables, and then counting the dots in each "cell."

Figure 10.4. Scatterplots for Positive and Negative Relationships

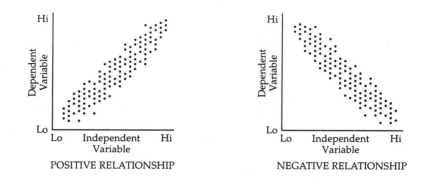

POSITIVE RELATIONSHIP   NEGATIVE RELATIONSHIP

Also recall ANOVA's box-and-whisker diagram in Section 9.1 showing the distribution of dependent variable scores within categories of the independent variable. As noted above, a box-and-whisker diagram is really just a special type of scatterplot—one with a categorized independent variable. Intriguing, isn't it, how statistical techniques are so closely related to one another. So much of statistics consists of variations on common themes. That's part of the beauty and elegance of statistics.

It is easy to see that scatterplots show the direction of a relationship. They also show a relationship's strength. If a relationship is perfect, the data points form a perfectly straight line. If two variables are strongly (but not perfectly) related, then scatterplot points cluster tightly around a straight line that we can imagine running through the data points. For a moderate relationship, points are more dispersed, although with a directional pattern still evident. For a weak relationship, scatterplot points are very widely dispersed and the direction of the relationship may be hard to identify. And for no relationship at all, points are randomly distributed about the scatterplot. Figure 10.5 illustrates each of these situations in somewhat exaggerated, ideal form.

Can you see why these patterns reflect a relationship's strength? A strong relationship is a predictable relationship in that we can estimate a dependent variable score very accurately if we know a case's independent variable score. If scatterplot data points are in a perfectly straight line, we can find a case's dependent variable score exactly if we know the case's score on the independent variable. All we do is locate the independent variable score on the X-axis, find the point on the line of data points directly above it, and then locate the height of that point on the Y-axis.

Figure 10.5. Scatterplots for Different Strengths of Relationships

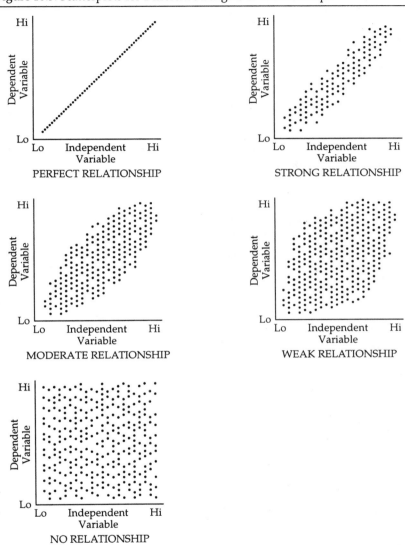

For weaker relationships, data points are scattered above any given point on the X-axis, making it more difficult to estimate a dependent variable score if we know an independent variable score. After all, not all cases with that score on the independent variable have the same dependent variable score. There is variation in dependent variable scores for cases with the same independent variable score. When points are randomly distributed about a scatterplot, estimation of dependent variable scores is random—just a wild guess. Then the two variables are not related.

In summary: For perfect relationships, the variance in dependent variable scores associated with a given independent variable score is 0. The variance increases as relationships weaken, and finally the variance approximates the variance of the total set of scores if the variables are unrelated.

Again we find a close analogy to cross-tabulation. A table for strongly related variables has cases clustering along a diagonal—the minor diagonal (/) for positive relationships and the major diagonal (\) for negative relationships. Weakly related variables have a table with cases largely off the diagonals and scattered about the cells. Likewise for patterns shown in scatterplots.

10.3   Some Limitations of Scatterplots

So a scatterplot shows both the direction and strength of a relationship. That's very helpful. But scatterplots sometimes have a practical problem. The two-dimensional space of a scatterplot makes it difficult to display data for a large number of cases. We have already seen this problem in ANOVA's box-and-whisker diagrams. If N is large and/or variable scores do not assume very many different values, scatterplot data points often pile up on top of one another and are hard to display on a two-dimensional graph. There are some ways to do so, the simplest using different symbols to show different numbers of cases located at points on the scatterplot. A triangle might stand for 1 to 5 cases, a square for 6 to 10 cases, and so on. But symbols like these do not help much; visualization is still difficult. There are also computer graphing techniques that "jiggle" coinciding data points to scatter them just a bit so we can see them or that visually add a third depth dimension showing the number of cases at each point, but such techniques are found only in more sophisticated computer programs.

This limitation of two-dimensional scatterplots can be quite serious. You can well imagine what happens with most scatterplots showing relationships between variables in a large data set like the General Social Survey with its hundreds of cases. With a large N, there are likely to be many cases with the same joint values of the independent and dependent variables. We can't tell how many cases each dot on a scatterplot represents. Even in our 50-case data set for only the most populous countries, a few scatterplot dots almost coincide (Russia, France, and Taiwan have nearly identical urban percentages of 73, 74, and 75, respectively, and exactly the same fertility rate of 1.8). Generally speaking, scatterplots are most useful when we have fewer than, say, 100 cases. In practice this means that scatterplots are usually more useful with aggregate data like those for nations of the world, American states, or Canadian provinces than with survey data like the General Social Survey since aggregate data sets often have relatively few cases.

But whatever its practical limitations with a large number of cases, a scatterplot is always useful conceptually. It does summarize a relationship visually. And we have seen that it works very well with a data set (like the 50-nations data) that has a fairly small number of cases.

10.4   Regression and Least-Squares Lines

A scatterplot is a good visual summary of a relationship. However, we can summarize a relationship even more concisely with a single straight line that "best" describes the relationship. This is the line I had in mind when I suggested we could imagine a line running through the data points on a scatterplot. In Figure 10.6 I have drawn a straight line that seems to capture the urbanism → fertility relationship pretty well.

Figure 10.6. Fertility Rate by Percent Urban (with Regression Line)

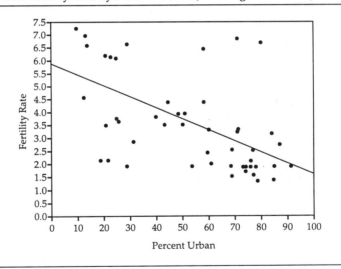

Yes, that single straight line seems to fit the general pattern of scatterplot dots quite closely. One country's data point lies exactly on the line, and a couple of countries come close and most don't lie too far from the line. The line is a pretty good fit of the data points. But why draw this *particular* line? Why not a line a little above the one I have drawn? Or a little below? Or at a slightly different angle? How do I know that this particular line is the *very best* straight line I could draw?

The answer is a little complicated: The line I have drawn minimizes the sum of the squared vertical distances between the line and the dependent variable score of each case. In other words, there is no other line for which the sum of the squared vertical distances to the dependent variable score of each case is less. In Figure 10.7 I have drawn vertical lines—called ***residuals***—showing the distance between each of our 50 cases and the straight line summarizing the urbanism → fertility relationship. If you were to measure the length of each of the 50 residuals, square those lengths, and then add up the squares, the sum you get would be less than you could get for any other line you could draw to summarize the urbanism → fertility relationship.[1]

Figure 10.7. Percent Urban by Fertility Rate (with Examples of Residuals)

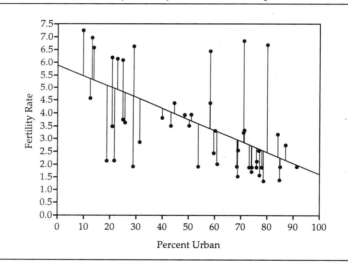

This summary line is called the ***regression line*** or ***least-squares line***. It is very important in statistics, and we will use it often in this chapter and Chapter 12 and throughout much of advanced statistics.

Why do we want the line that minimizes the sum of squared distances to dependent variable scores? Because this particular line—the regression line—best predicts a case's score on the dependent variable if we know the case's independent variable score. Consider: To predict a country's fertility rate knowing nothing else

---

[1] It is also true that if we treat vertical lines above the summary line as positive and vertical lines below the summary line as negative, then the sum of the lengths of all the vertical lines is 0. The total length of the lines above exactly cancels the total length of the lines below. But it is more important for us that the sum of the squares of the vertical line lengths is a minimum.

about the country, the best guess we can make is the mean fertility rate of all 50 countries. The reason the mean fertility rate is the best estimate is that, as we saw in Section 3.4, the mean minimizes the sum of squared distances from each score. In this important sense, no other number produces a better estimate of an individual case's score than the mean of all scores.

But if fertility is related to urbanism, then we can make use of a country's urbanism score to better predict or estimate its fertility rate. If (like Egypt) a country has an urbanism percentage of, say, 45, we find the fertility rate associated with an urban percentage of 45 on the regression line. Just eyeballing the scatterplot and its regression line in Figure 10.7, it looks like a country (like Egypt) that is 45 percent urban has a fertility rate somewhere around 4.

But 4 is just a rough visual estimate. We can be more precise if we recall (from high school math) that any straight line can be expressed as a simple equation with this form:[2]

$$Y = a + bX$$

where Y = score on the dependent variable

   a = **Y-intercept**, or the value of Y at which the line crosses the Y-axis[3] (i.e., the value of Y when X = 0)

   b = **slope**, or the change in Y for every one-unit increase in X (i.e., the steepness of the line)

   X = score on the independent variable

This equation is called the **regression equation** for the regression of fertility on urbanism. Note the wording: We speak of *dependent variable Y's regression on independent variable X*. Using a straight line to describe a relationship is called **linear regression**. By the way, sometimes the Y-intercept is called the **constant**.

We'll find out later in this chapter how to calculate the values of a and b in a regression equation, but for now let me just tell you that a = 5.720 and b = −.041 for the regression of fertility rate on percent urban. That is, the regression equation is

$$Y' = 5.720 + (-.041)X$$

---

[2] You may be used to representing a straight line with the equation $Y = mX + b$, with m = slope and b = Y-intercept. Clearly, however, this representation is exactly equivalent to $Y = a + bX$, differing only in the symbols used for slope and intercept and in the arrangement of terms.

[3] A line crosses the Y-axis at a only on graphs on which the X-axis begins at 0. Sometimes, however, it is convenient and efficient to begin the X-axis at some other value. In the workbook for this chapter we will find that is what MicroCase does unless the variable displayed on the X-axis has a score near 0.

or more simply

$$Y' = 5.720 - .041X$$

I have added a ' mark (called a prime) to the symbol Y to remind us that the regression equation can be used to predict the value of Y. (Y' is pronounced "Y prime.")

The value of slope b in a regression equation is extremely important. Called a *regression coefficient*, it tells us the increase or decrease in the dependent variable associated with an increase of one unit in the independent variable, with the sign indicating the direction of the change. In our urbanism → fertility example, each additional percent of a country's population living in urban areas is associated with a decrease of .041 in fertility rate (i.e., in children per woman). If urbanism goes up one percentage point, then the average number of children born to each woman drops .041.

I know .041 children doesn't sound like much. That's only about 1/25 of a kid. But do the math: If a country's urbanism increases 25 percentage points (from, say, 45 to 70 percent), we expect a woman on average to have one fewer child. That is a sizable decline in fertility (and a lot fewer people in a country with tens or even hundreds of millions of women). Behind what may seem like dry statistics are real people living out their real lives and having—or not having—real children. In the end there is nothing abstract at all about statistical analyses.

Take Egypt as an example. We predict Egypt's fertility rate by plugging its urban percentage into the urbanism → fertility regression equation, like this:

$$Y' = 5.720 - .041\ (45)$$

$$= 5.720 - 1.845$$

$$= 3.875$$

Egypt's predicted fertility—predicted on the basis of its urbanism—is 3.88 children per woman.

We certainly make errors in such predictions. After all, the regression line does not go through each and every data point. Most cases lie somewhere off the regression line. We have just seen a case, Egypt, with an actual or observed fertility rate of 4.35. Were we predicting Egypt's fertility rate on the basis of our regression line or equation, we would make an error of 4.35 − 3.88 = .47. That's the difference between Egypt's actual fertility rate and predicted rate. Egyptian women have about .47 more children than we would expect on the basis of the percentage of its population who live in urban areas.

But although our predictions are not exactly right, on average we cannot do any better than to estimate fertility rates using the regression line and its equation because the regression line minimizes the sum of squared errors. After all, the distance between a case's dependent variable score and the regression line is error. More specifically, the distance measures error in predicting the dependent variable score from knowledge of the independent variable score for that case. To measure total error for all cases, we square all the individual errors and then add those squared errors, just as we did when computing the sum of squares for the numerator of the variance in Section 4.1.[4] The less error, the smaller this sum of squared distances; the more error, the larger this sum. So, if we make predictions for all 50 cases, then square the difference between each actual fertility score and predicted fertility score, and finally sum all the squared differences, we would find a smaller sum than for any other straight line we could possibly draw through our scatterplot dots.

Here, then, is the punch line of all this regression-line business: The regression line minimizes the sum of the squared errors and is thus the best possible straight-line predictor of dependent variable scores.

10.5   Calculating Regression Coefficients

We have seen that a linear relationship can be described with the equation

$$Y = a + bX$$

Now we need to find how to calculate the values of a and b in this equation. Let's take the regression coefficient b first. The formula for b is:

$$b = \frac{\Sigma(X - \overline{X})(Y - \overline{Y})}{\Sigma(X - \overline{X})^2}$$

The numerator is found by subtracting the variable's mean from each score, multiplying these differences for the independent and dependent variable scores for each case, and then adding the products. The product of the deviation of scores from their means [that is, $(X - \overline{X})(Y - \overline{Y})$ for each case] is called the *cross-product*. Therefore, the numerator in the formula for b is the sum of cross-products.

We have seen the denominator in the equation for b several times before. It is our old friend, the sum of squares of the independent variable X. Although we don't need to do so here, if we divide $\Sigma(X - \overline{X})^2$ by $N - 1$, we have the estimate of the variance of X. We

---

[4] It won't do to just add all the unsquared errors since that sum is always 0. We saw the parallel of this situation in Section 4.5 where we found that the sum of differences between scores and the mean is always 0.

found this sum of squares and this variance first in Chapter 4 and again in Chapter 9. Now we find it once again.

The regression coefficient b, therefore, is the ratio of the sum of cross-products to the sum of squares of the independent variable. The regression coefficient is a useful measure of the effect of the independent variable on the dependent variable since it tells us how much change in the dependent variable is associated with one unit increase in the independent variable. Of course, the magnitude of the regression coefficient depends on our units of measurement. The regression coefficient is larger, for example, if variables are measured in inches rather than feet or in minutes rather than hours. Nevertheless, if we keep our measurement units in mind, the regression coefficient is a useful gauge of the effect of an independent variable on a dependent variable.

To find the intercept a in the regression equation, note that we can rearrange the regression equation terms as

$$a = Y - bX$$

Now let's make use of an interesting property of the regression equation: It always goes through the point at which the X and Y means cross. If we draw a line at the X mean and a line at the Y mean on a scatterplot, the point at which the lines intersect is on the regression line. This is always true. Thus, we know that the joint values of $X = \overline{X}$ and $Y = \overline{Y}$ satisfy the equation for a, and thus $a = \overline{Y} - b\overline{X}$. We find a by entering the values for b and the means $\overline{Y}$ and $\overline{X}$ in this equation. For the regression of fertility on urbanism, b is −.041, the fertility mean is 3.563, and the urbanism mean is 52.140 (unrounded), giving

$$
\begin{aligned}
a &= \overline{Y} - b\overline{X} \\
&= 3.563 - (-.041)\,52.140 \\
&= 3.563 + 2.138 \\
&= 5.701
\end{aligned}
$$

Entering values of the intercept a and slope b into the regression equation, we have

$$Y' = 5.720 - .041X$$

10.6   Correlation Coefficient (r)

We observed earlier that the strength of a relationship is reflected in how tightly scatterplot dots cluster around the best-fitting straight line. We now know that a straight line is the regression line that minimizes the sum of squared distances to cases' dependent variable scores. If the relationship is strong, then cases cluster closely about the regression line. If the relationship is weak, cases are widely distributed about the regression line. If there is no relationship, then cases are randomly distributed in the scatterplot and thus very dispersed from the regression line.

What we need to assess the strength of a relationship is a good, concise measure of how tightly points cluster about a regression line. There is such a measure of association: *Pearson's product-moment correlation coefficient*. This lengthy moniker is usually shortened to *correlation coefficient*[5] and is symbolized *r*. Developed in the nineteenth century by Sir Francis Galton (who was Charles Darwin's cousin) and elaborated upon by the great British statistician Karl Pearson, the correlation coefficient is the most widely used measure of association in the social sciences.

The correlation coefficient r is a summary measure of how tightly cases are clustered around the regression line. If cases bunch very closely along the regression line, r is large in magnitude—indicating a strong relationship. If cases are widely dispersed about the regression line, then r is small in magnitude—indicating a weak relationship. Like well-behaved measures of association for ordinal variables (e.g., gamma), r ranges between –1.00 and +1.00 with the + and – signs indicating direction of relationship. As with ordinal measures of association, we usually omit the + sign from r unless there is some ambiguity about the direction of the relationship. r = –1.00 and r = 1.00 indicate perfect relationships, with the sign denoting direction. r = 0 means that the two variables are statistically unassociated—there is no relationship. The magnitude of r between 0 and ±1.00 reflects the strength of the association. By the way, a correlation coefficient for population data is sometimes symbolized ρ *(rho)*.

To understand correlation coefficients, let's think about strong and weak relationships. For brevity, I will focus only on positive relationships, but exactly the same logic also applies to negative relationships. If two variables are positively related, then cases with high scores on one variable generally have high scores on the other. And, too, cases with average scores on one variable by and large have average scores on the other. And, yes, cases with low scores on one variable tend to have low scores on the other. Obviously some cases may be exceptions to these patterns, but they will hold true for most cases.

---

[5] Strictly speaking, "correlation coefficient" is a generic term embracing a variety of measures of association. However, the full name, Pearson's (or Pearsonian) product-moment correlation coefficient, is usually used only if there may be some confusion with other kinds of correlation coefficients.

Now suppose that instead of using raw scores, we convert all scores into standard scores—that is, into Z-scores. We learned how to do this in Section 4.3. Z-scores standardize scores so that they are not affected by the unit of measurement. This is good since we certainly don't want a measure of association to be affected by whether variables use inches or feet, ounces or pounds, dollars or cents.

Remember (from Section 4.3) that we convert a score into a standard score by subtracting the mean and then dividing the result by the standard deviation. For each case, suppose we convert both its independent variable score and dependent variable score into Z-scores. This operation changes the scale of the scores, but it does not change their positions relative to the scores of other cases. High scores will still be relatively high, average scores still average, low scores still relatively low. All calculation of standard scores does is to give each variable the same scale. After standardization, both the independent and dependent variables have means of 0 and standard deviations of 1.00, just as all standard scores do.

Since we have changed only the variables' scales, use of standard scores does not change the nature of a relationship. In a strong positive relationship, low Z-scores on the dependent variable generally are associated with low Z-scores on the independent variable. And high Z-scores on the dependent variable usually accompany high Z-scores on the independent variable. If two variables are *perfectly* related, then each case's Z-score on the dependent variable will exactly equal its Z-score on the independent variable. For a perfect relationship, then, $Z_X = Z_Y$.

Researchers often denote variables with subscripts, with the first subscript denoting the dependent variable and the second subscript denoting the independent variable. Thus, r_{YX} is the correlation coefficient for the relationship between dependent variable Y and independent variable X.

Now we are ready for a formula for the correlation coefficient.[6] For population data:

$$r = \frac{\Sigma Z_X Z_Y}{N}$$

where r = Pearson product-moment correlation coefficient

Z_X = standard score on the independent variable X

Z_Y = standard score on the dependent variable Y

N = number of cases

---

[6] In the days before computers, computational formulas rather than this definitional formula were usually used to calculate correlation coefficients. We have no need for such computational shortcuts now that we have computers. Consult a traditional statistics text if you want computational formulas.

$Z_X Z_Y$ is the cross-product of scores in standard form. This formula tells us that Pearson's correlation coefficient is the average cross-product of scores in standard form. The formula given here is for population data like the data for the 50 most populous nations. For sample data like the General Social Survey, we divide $Z_X Z_Y$ by $N - 1$ rather than N. Whether we use N or $N - 1$ makes little difference, of course, when N is large.

We are interested in r because it has marvelously useful interpretations. Think back to Section 4.3. (In fact, reread 4.3 if you are a little rusty on standard scores.) Remember that Z-scores have the interesting property that $\Sigma Z^2 = N$. Well then, if two variables are perfectly related, their Z-scores are the same (i.e., $Z_X = Z_Y$), and thus

$$\Sigma Z_X Z_Y = \Sigma Z_X^2 = \Sigma Z_Y^2 = N$$

And this in turn means that

$$r = \frac{\Sigma Z_X Z_Y}{N} = \frac{N}{N} = 1.00$$

So, if a relationship is perfect, $r = 1.00$. How convenient!

Note too that if two variables are unrelated, the products of their Z-scores cancel out when added up. That's because about half the products are positive and half the products are negative. (Remember that the mean of Z-scores is 0, so about half the scores are negative. Remember also that the product of a positive and a negative number is negative.) Thus, for unrelated variables, $\Sigma Z_X Z_Y = 0$ and therefore

$$r = \frac{\Sigma Z_X Z_Y}{N} = \frac{0}{N} = 0$$

So, if there is no relationship, $r = 0$. And for relationships greater than 0 but less than perfect, the magnitude of r indicates the strength of the relationship. Again, how convenient!

I have considered only positive relationships here, but exactly the same argument applies for negative relationships. For negative relationships, there will be more high scores paired with low scores than either high-high or low-low pairs. That, by definition, is the pattern that makes for a negative relationship. Therefore, $\Sigma Z_X Z_Y$ will be negative and, therefore, r will be negative.

Note from the formula for the correlation coefficient that r is a symmetric measure of association. It does not matter which of the two variables is dependent and which is independent. After all, the order in which Z_X and Z_Y are multiplied doesn't matter because $Z_X Z_Y = Z_Y Z_X$. The correlation between them is the same either way. Using subscripts to designate variables (with the dependent variable always denoted first), $r_{YX} = r_{XY}$.

Incidentally, r for the relationship between urbanism and fertility is –.56, a fairly strong relationship. Calculation of this correlation is drudgery, so I won't work through the computations involved, although I welcome you to do so. To do so, first calculate the means and standard deviations for the urbanism and fertility scores. Then use those means and standard deviations to convert each country's scores into standard scores, multiply each country's standard scores together, and add all those cross-products. Finally, divide the total of cross-products by the number of countries to get the correlation coefficient. Yes, I'll leave those calculations entirely to you. (Actually, there is a computational formula that was handy before computers, but the formula isn't worth bothering with these days.)

Francis Galton and Karl Pearson were ingenious. They created a correlation coefficient for interval/ratio variables that varies from –1.00 for a perfect negative relationship to 0 for no relationship to +1.00 for a perfect positive relationship. The sign indicates the direction, and the magnitude gauges the strength of a relationship. The correlation coefficient r thus behaves like the cross-tabulation measures of association for ordinal variables that we learned in Chapter 7.

As with cross-tabulation's percentage differences and measures of association, differences between means, and eta squared in ANOVA, we need to assess the magnitude of any r in the light of our theory, expectations, and findings from other studies. There are no hard and fast rules for saying that a certain r indicates a strong relationship or a particular r indicates a moderate relationship. The world is much too continuous and subtle for such absolute rules. But as a general guideline, to be tempered by your own thinking and good judgment, you can use the following *very rough* equivalents to assess correlation coefficients:

Negative Relationship				No Relationship			Positive Relationship			
r = –1.00	–.80	–.60	–.40	–20	.00	.20	.40	.60	.80	1.00
Perfect	Strong	Moderate	Weak	None		Weak	Moderate	Strong	Perfect	

I have described the correlation coefficient and the regression model on which it is based as appropriate for interval/ratio variables. That's true. But it is also true that most researchers readily extend regression and correlation techniques to ordinal variables. This practice is analogous to what we found for means. Recall from Section 4.3 that, strictly speaking, means are calculated only for interval/ratio variables, but researchers often find it helpful to use means for ordinal data too. Likewise, researchers often compute correlation coefficients for ordinal variables when doing so helps identify patterns in data. In Chapter 12 we will learn a technique using dummy variables for incorporating even nominal variables into regression and correlation analyses.

10.7   r² as Proportion of Variation Explained

We can interpret the square of r as a proportional reduction in error (PRE) measure of association. That is, r^2 reports the proportion that we can reduce errors in predicting cases' dependent variable scores if we know their independent variable scores. Let me explain why.

I pointed out in Section 3.4 that the mean is the value that minimizes the sum of squared differences between itself and each score. Since the mean minimizes the sum of squares, we can do no better than to guess the mean if we want to guess cases' scores on the dependent variable. We won't often be exactly right, of course, but we will minimize our errors. The sum of squares measures the amount by which we are wrong.

But what if we use our knowledge of independent variable scores to guess scores on the dependent variable? We have seen that if two variables are related, knowledge of the independent variable reduces errors in predicting dependent variable scores. We can enter independent variable scores into the regression equation to estimate dependent variable scores. We will still make errors, but the errors will be smaller than if we didn't use our knowledge of independent variable scores. The errors that remain are residuals—expressed graphically, as we have seen, by the vertical distances between data points and the regression line. "Residuals" is an apt name because they are "left over" in that they are left unexplained by the independent variable. We can measure the total error by the sum of the squares of these residuals.

The reduction in errors—that is, errors not using independent variable scores minus errors using independent variable scores—is the amount that the independent variable has helped us explain dependent variable scores. We can measure the amount of this explanation. It is the total variation in the dependent variable minus the variation that is left after taking the independent variable scores into account. That's how much of the variation in the dependent variable we have explained by using our knowledge of independent variable scores.

We usually express this explained variation as a proportion of total variation—easily done by dividing the explained variation by the total variation. Now the payoff for thinking about all this: The correlation coefficient squared—r^2—is this ratio of variation explained by the independent variable to total variation. Algebraically:

$$r^2 = \frac{\text{Explained Variation}}{\text{Total Variation}}$$

This ratio is sometimes called the *coefficient of determination*. For the urbanism → fertility relationship, $r^2 = (-.56)^2 = .31$. Thus, percent urban explains 31 percent of the variation in fertility among the 50 most populous countries.

Interpretation of r^2 as the proportion of variation explained should sound familiar. It is directly analogous to the interpretation of E^2 described in Section 9.5 on analysis of variance. The difference is that r^2 assumes a linear (i.e., straight line) relationship whereas E^2 makes no assumption about the form of the relationship. E^2 assumes that we estimate dependent variable scores using the dependent variable means within categories of the independent variable. If a relationship is truly linear and thus these means are arrayed in a perfectly straight line, r^2 equals E^2. Otherwise (and this is almost always the case), E^2 is larger than r^2.

Since r^2 is the proportion of variation explained, and since total variation equals explained variation plus unexplained variation, it follows that $1 - r^2$ is the proportion of variation that is unexplained by the independent variable. For example, about 69 percent (i.e., $1 - .31 = .69$) of the variation in fertility is unexplained by variation in urbanism among the 50 most populous nations. This statistic, $1 - r^2$, is called the *coefficient of alienation*.

Keep in mind when interpreting either r or r^2 for ecological data that you cannot infer relationships among variables at the level of individuals. Finding a strong relationship among countries does not necessarily mean that urban women have fewer children than rural women. Maybe they do . . . and maybe they don't. We would need individual-level data, not ecological data, to find out and avoid an ecological fallacy. But we do know that countries with more urbanism tend to have fewer children per woman.

10.8   Correlations Between Dichotomous Variables

Regression and correlation techniques assume that variables are measured at the interval/ratio level. Most researchers are also willing to extend these techniques to ordinal variables. But regression and correlation can be extended even to dichotomous variables. Let me explain.

I pointed out in Section 3.4 that the mean has an interesting and useful characteristic for dichotomous variables: If a variable's values are coded 0 and 1, then the variable's mean is the proportion of cases with the characteristic distinguished by code 1. Take as an example the dichotomous variable sex, with males coded 0 and females coded

1.[7] The mean of sex (.57 in the General Social Survey) is the proportion female (i.e., the proportion of cases coded 1).

But here is another interesting and useful correspondence between two statistics: The correlation coefficient r between two dichotomous variables is equal to ϕ (phi), the nominal-level measure of association for 2 by 2 tables that we met in Section 7.2. And since $\phi = r$, it follows that $r^2 = \phi^2$, so ϕ^2 reports the proportion of variation in one dichotomous variable explained by variation in another dichotomous variable. If (as in the General Social Survey) the correlation between sex and fear of walking in one's neighborhood is .29, then sex explains 8 percent [$(.29)^2 = .08$] of the variation in respondents' fear of walking in their neighborhoods.

Yes, with $\phi^2 = r^2$, we see once again that statistics is a fine-meshed web of interrelated ideas and techniques. This equivalence of ϕ^2 and r^2 is the basis for interpreting ϕ^2 as a PRE (proportional reduction in error) measure of association and thus expands the usefulness of ϕ. Do keep this in mind whenever you are working with dichotomous variables, even ones measured at the nominal level. ϕ^2 is a handy statistical tool in such situations. We'll learn in Chapter 12 to extend regression and correlation to even *nondichotomous* nominal-level variables, but for now restrict regression and correlation to interval/ratio variables and dichotomous variables, and maybe, with caution, to ordinal variables.

10.9 Association *Still* Does Not Imply Causation

A brief but very important reminder: Statistical association does not imply causation. Never has/never will. Two variables might be correlated even though there is no causal connection between them. You have heard this before. Several times, in fact. I pointed out that association does not imply causation with regard to cross-tabulation, the difference between means, or ANOVA. The same observation applies to correlation coefficients. Just because two variables are correlated—even highly correlated—does not mean that one causes the other. We know that urbanism and fertility are associated with one

---

[7] In fact, males are coded 1 and females are coded 2 in the General Social Survey. Codes using consecutive numbers like 1 and 2 rather than 0 and 1 have no effect on the logic of dummy variables or on regression and correlation coefficients. The only effect is increasing the Y-intercept in the regression equation and, of course, the mean of the variable. However, with 0 and 1 codes, the mean equals the percentage of scores with code 1—a convenient equivalence. Moreover, 0 and 1 are easier to conceptualize as dummy variable codes since 0 symbolizes the absence and 1 the presence of whatever is measured (here, "femaleness"). Therefore, I am using 0 and 1 as dummy codes in this example.

another, so maybe urbanism affects fertility rates. But we can't be sure of a causal connection. Remember the storks and babies, spelling scores and foot sizes . . . . We will look at causation more fully in Chapters 11 and 12.

10.10   Linear and Nonlinear Relationships

In describing regression and correlation coefficients, we assumed that the regression line is a straight line. That is, in fact, what we usually assume. The reason is that straight lines are both conceptually and mathematically simpler than other lines. Since a major purpose of descriptive statistics is reduction of information to make it more manageable, we find straight lines the easiest way to summarize relationships. And this approach often works fine since many relationships are described remarkably well with straight lines. This approach is called linear regression. That is what we have been doing in this chapter.

But not all relationships are linear. Figure 10.8 depicts some scatterplots showing obvious relationships, but ones that cannot be summarized very well (or even at all) with straight lines. These are curvilinear relationships. We have seen them before with cross-tabulation data in Section 5.4. Linear regression and correlation techniques are inadequate to handle such patterns. The correlation coefficient r underestimates the strength of these relationships. In fact, for some curvilinear relationships like these, r is 0 or close to it, indicating—incorrectly—that the two variables are not related. They certainly are related, but not in a linear fashion.

You should watch for nonlinear patterns. Although curvilinear relationships are rarely as pronounced as those shown in Figure 10.8, you can usually recognize them on scatterplots unless the number of cases is so large that it obscures interpretation of a scatterplot. You should, therefore, begin any regression and correlation analysis by examining a scatterplot along with the distributions of individual variables.

Figure 10.8. Scatterplots of Curvilinear Relationships

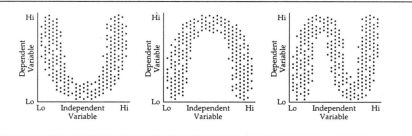

If relationships appear nonlinear, then linear regression and correlation techniques should not be used. There are advanced statistical techniques that handle nonlinear relationships by mathematically "straightening" them with transformations. Sometimes we "straighten" relationships by using the logarithm of scores rather than raw scores. Other times the square of raw scores or the sine of raw scores work better. There are in fact many transformations available, the best in a given situation depending on the distribution of the variables. These transformations are a little beyond what we can cover in this text, but I want you to know that they exist.

10.11   Test of Significance for a Correlation Coefficient

It is possible that we will find a nonzero correlation coefficient in sample data even if the "true" correlation coefficient in the population is 0. After all, even if there is no relationship between two variables in a population, just by chance we could randomly select a sample that yields a nonzero positive or negative correlation. In Chapter 6 we found that the chi-square test of statistical significance assesses this possibility for cross-tabulations. In Chapter 8 we applied the t test to gauge this possibility. In Chapter 9 we learned that the F test evaluates this possibility in an analysis of variance. And yes, here we will learn about a test of significance that estimates the probability that a given correlation coefficient occurs only by chance if there is no population relationship.

This test of significance for r requires that we make a few assumptions. First, we assume a linear relationship in the population. Second, we assume that we have taken a random sample of the population. Third, we assume that the two variables are normally distributed in the population. Violation of this last assumption, however, has serious consequences only when we have a small number of cases—say, fewer than 30. When we have 30 or more cases, we can relax the assumption that our variables are normally distributed in the population.

If we can make these assumptions, we carry out the significance test by computing an F ratio using the following formula:

$$F = \frac{r^2(N-2)}{1-r^2}$$

where r = correlation coefficient

N = number of cases

The degrees of freedom for this test are 1 and N – 2. After we compute the F ratio and its degrees of freedom, we determine the level of significance from Appendix Table 3. We find 1 degree of freedom in the first column of F values. N – 2 are the degrees of freedom described in the rows of the F tables. If F is larger than the value we find in the F table in the Appendix, then we reject the null hypothesis of no relationship between the two variables and conclude that the two variables are probably related to each other in the larger population from which the sample was drawn.

But we often carry out significance tests on correlation coefficients even for population data. As noted in Section 6.5, significance tests allow us to assess the likelihood that a relationship in population data is due to random processes. Here is the F ratio for the urbanism $\rightarrow$ fertility relationship ($r = -.56$) among the 50 most populous nations:

$$F = \frac{r^2(N-2)}{1-r^2}$$

$$= \frac{(-.56)^2(50-2)}{1-(-.56)^2}$$

$$= \frac{(.314)(48)}{1-.314}$$

$$= \frac{15.07}{.686}$$

$$= 21.968$$

Table 3 in the Appendix indicates that with an F of 21.968 and 1 and 48 degrees of freedom, r of –.56 is significant at the .001 level. There is less than one chance in a thousand that we would find a correlation this large or larger if there is "really" no relationship. Those are pretty small odds, so we reject the null hypothesis of no relationship. We conclude that there probably is a relationship between urbanism and fertility, and that it is not explained by random processes. If our data were from a sample rather than a population, we would feel confident in generalizing from our sample data to the population from which it was drawn and in concluding that the variables are also related in the population.

10.12   Correlation Matrix

Rarely do we work with just two variables in social scientific analyses. Even in bivariate analyses, we typically are interested in interrelationships among a larger set of variables. A *correlation matrix* is a convenient format for presenting correlation coefficients describing interrelationships among three or more variables. Table 10.2 presents a correlation matrix showing bivariate relationships among a set of four variables. I have included number of radios per 100 persons and gross domestic product (GDP) per capita along with fertility rate and percent urban. Each cell above the \ diagonal reports the correlation between the row variable and column variable.

Table 10.2. Correlations Among Fertility Rate and Three
Independent Variables (pairwise deletion)

Variable	Fertility Rate	Percent Urban	Radios per 100	GDP per Capita
Fertility Rate	1.00	–.56**	–.47**	–.61**
Percent Urban	50	1.00	–.62**	–.67**
Radios per 100	49	49	1.00	.75**
GDP per Capita	39	39	39	1.00

** $p < .01$

Correlations in the major diagonal are always 1.00 because variables are always perfectly correlated with themselves. Since a correlation coefficient is a symmetric measure of association, a correlation matrix does not need to repeat the correlation coefficients below the diagonal. The triangle of correlations below the diagonal would be a mirror image of the triangle above. Here these lower cells in the matrix report the number of cases on which each correlation is based—a good practice if correlations are based on different Ns. Sometimes significance levels, (e.g., .05, .01, or .001) of correlation coefficients are presented in corresponding cells below the diagonal. I have not done so here, but instead indicate significance levels with asterisks keyed to the probability footnoted below the matrix.

The varying Ns mean that some correlations in Table 10.2 are based on somewhat different sets of cases than are other correlations. Each bivariate correlation in Table 10.2 is based on whatever data is available for the two variables being correlated. Complete data is available for fertility rate and percent urban, and the correlation between these two variables uses all 50 cases. Uzbekistan, however, lacks information on radios per 100 persons, and is necessarily excluded for correlations involving radio availability. More seriously, the three correlations involving GDP per capita are able to use data for only 39 cases. The other 11 cases (which include Uzbekistan) are missing GDP data. For example, Canada is one of the 11 countries missing GDP information. Canada is necessarily excluded from the three correlation coefficients involving this variable. But Canada is included in the three correlations that do not involve GDP.

This method of handling missing data—exclusion only of cases missing information on the particular variable used in the calculation of a given statistic—is called *pairwise deletion*. An alternative method of handling missing data—*listwise deletion*—excludes any case that is missing information on any variable in the entire analysis, even if the case has complete data for a given pair of variables. Table 10.3 presents correlations among the same set of variables but with listwise deletion used to handle missing data. I've abridged the matrix by omitting the diagonal and lower cells. I could have used these cells to report significance levels, but I have instead indicated significance levels with asterisks. Besides, I want you to see a variation on the format for a correlation matrix.

Table 10.3. Correlations Among Fertility Rate and Three Independent Variables (listwise deletion, N = 39)

Variable	Percent Urban	Radios per 100	GDP per Capita
Fertility Rate	−.64**	−.46**	−.61**
Percent Urban		.61**	.67**
Radios per 100			.75**

** p < .01

Analogous correlations in Tables 10.2 and 10.3 are far from identical. Uzbekistan, Canada, and the nine other countries missing information on the GDP are excluded from each correlation in Table 10.3 even though the data set has information on the fertility rate, percent urban, and radios per 100 people for each of these countries. Thus, the two sets of correlations are based on different sets of cases. Compare Tables 10.2 and 10.3 and you will see that pairwise and listwise deletion may produce quite different statistical results, especially when a significant proportion of cases included by pairwise deletion are excluded by listwise deletion. Pairwise deletion has the advantage of making maximum use of whatever data are available. Listwise deletion has the advantage of basing all statistics (e.g., correlation coefficients) on exactly the same set of cases. We will find more examples of listwise deletion when we consider multivariate analyses in Chapters 11 and 12.

Tables 10.2 and 10.3 are correlation matrices involving only four variables, but you can easily imagine a larger matrix with more variables. Be careful what you call an analysis displayed in this sort of correlation matrix. Even if three or more variables are displayed, the analysis that the matrix reports is still bivariate, not multivariate, because each statistic—in this example, each correlation coefficient —describes the relationship between only two variables. Tables 10.2 and 10.3 present correlation coefficients describing six different bivariate relationships and, therefore, report bivariate analyses.

10.13   Summing Up Chapter 10

Here is what we have learned in this chapter:

- Scatterplots display bivariate relationships between interval/ratio variables.
- Scatterplots are easier to interpret if there are relatively few cases.
- A least-squares regression line minimizes the sum of squared deviations of dependent variable scores from the regression line.
- The regression coefficient (the slope of the regression line) describes the change in the dependent variable for every increase of one unit in the independent variable. The regression coefficient thus measures the effect of the independent variable on the dependent variable.
- The correlation coefficient r is a measure of association for interval/ratio variables that varies between −1.00 and +1.00, with the sign indicating the direction of the relationship.
- The correlation coefficient r is the average cross-product of Z-scores.
- The square of a correlation coefficient (r^2) describes the proportion of the total variation in the dependent variable explained by the independent variable.
- For the relationship between two dichotomous variables, $r = \phi$ and $r^2 = \phi^2$. Thus, ϕ^2 reports the proportion of variation in one variable explained by the other variable.
- Linear regression and correlation assume straight-line relationships and are inappropriate for describing curvilinear relationships.
- Just because two variables are correlated does not mean they are causally related.
- The significance test for r requires assumptions of a linear relationship, random sampling, and normally distributed variables.
- We can relax the assumption of normality for significance tests of r if N is 30 or larger.

- A correlation matrix is a convenient format for presenting bivariate correlation coefficients and their significance levels.
- Two methods for handling missing data are available for bivariate analyses: pairwise deletion that makes use of all available information and listwise deletion that includes only cases with information on all variables in an analysis.

Key Concepts and Procedures

Ideas and Terms

scatterplot (or scattergram)	constant
X-axis	regression coefficient
Y-axis	cross-product
origin	correlation coefficient (Pearson's pro-
residual	duct-moment correlation coefficient)
regression line or least-squares line	coefficient of determination
Y-intercept	coefficient of alienation
slope	pairwise deletion
regression equation	listwise deletion
linear regression	correlation matrix

Symbols

Y'

a

b

r (r_{YX})

r^2 $(r_{YX})^2$

ρ

F

Formulas

$$Y' = a + bX$$

$$b = \frac{\Sigma(X - \overline{X})(Y - \overline{Y})}{\Sigma(X - \overline{X})^2}$$

$$a = \overline{Y} - b\overline{X}$$

$$r = \frac{\Sigma Z_X Z_Y}{N}$$

$$r^2 = \frac{\text{Explained Variation}}{\text{Total Variation}}$$

$$F = \frac{r^2(N - 2)}{1 - r^2}$$

ANALYSIS WRITE-UP 6: REGRESSION AND CORRELATION

✦ ✦ ✦ ✦ ✦

A. *Here is a write-up that presents a scatterplot along with unstandardized regression and correlation coefficients.*

Social scientists often measure occupational prestige with a scale ranging from 0 to 100. Previous studies have found strong correlations between occupational prestige and education. Figure 1 describes the education-prestige relationship for 67 General Social Survey respondents who report their ethnicity as Mexican. Education has a strong positive effect on occupational prestige among Mexican-Americans. The unstandardized regression coefficient of 2.17 indicates a substantial "payoff" of well over two points in occupational prestige for each additional year of education. The correlation coefficient of .55 is statistically significant at the .01 level. About 30 percent of the variation in occupational prestige among Mexican-Americans is accounted for by their years of education.

FIGURE 1 ABOUT HERE

Include figure at the end of the paper or report:

Figure 1. Scatterplot for Prestige by Education (Subset: Mexican Americans)

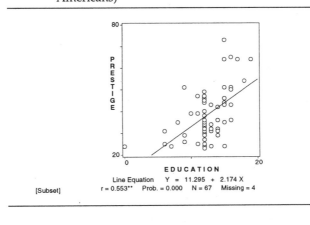

Line Equation   Y = 11.295 + 2.174 X
[Subset]   r = 0.553**   Prob. = 0.000   N = 67   Missing = 4

B. *It is not usually convenient to present scatterplots for numerous relationships, but regression equations and correlations can be efficiently summarized in tables. Here's an example.*

Table 1 presents unstandardized regression coefficients along with correlation coefficients describing the relationships between occupational prestige and education for four Hispanic ethnic groups distinguished in the General Social Survey. The regressions of occupational prestige on education have very similar least-squares lines for respondents who trace their roots to Spain or Mexico, and in each case education accounts for about 30 percent of the variation in prestige. Education and prestige are only slightly less correlated for the more heterogeneous category of "other Spanish." In all three of these Hispanic groups, education has a significant payoff in occupational prestige—between 1.75 and 2.5 prestige points for each additional year of education. However, this education-prestige relationship does not hold for respondents with Puerto Rican roots. Indeed, for Puerto Ricans, prestige and education have a weakly negative and statistically insignificant relationship.

TABLE 1 ABOUT HERE

Include table at the end of the paper or report:

Table 1. Regression and Correlation Coefficients for Occupational Prestige by Education, Hispanic Groups Only

Group	Intercept	b	r	p	N
Spaniards	8.403	2.57	.54	.001	34
Mexicans	11.295	2.17	.55	.001	67
Puerto Ricans	35.259	−.29	−.10	n.s.	21
Other Spanish	13.863	1.75	.50	.01	24

PART
III
Multivariate Analyses

CHAPTER 11
Multivariate Cross-tabulation

At the beginning of Chapter 5, I suggested six questions to address when analyzing a relationship. We have learned several ways to answer the first four: Is there a relationship in the sample? How strong is the relationship? What is the direction and shape of the relationship? And is there a relationship in the population? To answer these questions in a cross-tabulation, we compare and examine percentages, interpret measures of association, examine patterns among percentages, and carry out a chi-square test of statistical significance.

Now we will address the last two questions: Is the relationship genuinely causal, or is it instead a spurious relationship produced by some third variable? And if the relationship is genuinely causal, what intervening variable links the independent and dependent variables and thus accounts for the connection between them? Answers to these two questions require us to bring an additional variable (or more) into the analysis. In this chapter we learn to answer these questions with tabular techniques.

After this chapter you will be able to:

1. Explain the conditions that define a causal relationship.

2. Recognize and explain antecedent variables.

3. Recognize and describe causal explanation/spurious relationships.

4. Recognize and describe replication/genuine relationships.

5. Recognize and describe suppressor variables.

6. Explain the logic of introducing control variables (table elaboration).

7. Recognize and explain interpretation and intervening variables in causal relationships.

8.   Create and describe multivariate tables.

9.   Calculate and interpret partial gammas.

10. Describe the relationship between multivariate analysis and experimental design.

11.1   The Logic of Causal Relationships

We are so used to seeing the world in causal terms that few of us ever think much about just what a cause-and-effect relationship is. However much we may argue about whether some particular association is causal or not, we take the concept of causality itself pretty much for granted. But what exactly does it mean to say that smoking "causes" cancer? Or that poverty "causes" crime? Or that education "causes" attitude toward civil disobedience? In general terms, what does it mean to say that some independent variable X causes some dependent variable Y?

There is much agreement among both researchers and philosophers of science that we can say that variable X causes variable Y if, and only if, three conditions are met:

a.   The independent variable X must occur before the dependent variable Y.

b.   Variables X and Y must be associated with one another.

c.   The association between variables X and Y must not be due to some third, antecedent variable.

Let's consider each of these three conditions of causality in turn.

First, the independent variable X must occur before the dependent variable Y. In other words, the cause must precede the effect. Otherwise, the world would run backward, and that would be strange if not downright confusing. This is why movies run in reverse make us laugh. Cars uncrashing, bodies falling upward, or smoke pouring back into a gun barrel violate our sense of the proper temporal ordering of causal relationships. Movies like *Back to the Future* and its sequels convey the strangeness of a world in which the temporal ordering of causally connected events is violated.

Determining whether or not a relationship satisfies this first condition of causality is crucial, but it is not a statistical problem. Temporal ordering of variables is an issue for social theory and the methodology of research but not for statistical analysis. Plain common sense, our paradigm or worldview, or our theories (in the broadest of senses) may tell us that variables are ordered in a particular way so that we know, by "conventional wisdom" or our scientific theories, that the independent variable occurs before the dependent variable. It only "makes sense," for example, that most adults have completed their educations before answering a survey question about civil disobedience. Or research may be set up in a way that ensures the proper ordering of the independent and dependent variables. Experiments are designed so that the independent variable is manipulated before the researcher observes the effect of that manipulation on the dependent variable. Whether resolved by common sense, paradigms, theories, or research design, this temporal-ordering condition is not at issue in the statistical analysis of data.[1]

The second condition of causality: Variables X and Y must be associated with one another. That is, certain values of the dependent variable must be found occurring with certain values of the independent variable more often than we would expect by chance. Cancer rates must be higher among smokers, crime rates must be higher in poorer areas, support for civil disobedience must be higher among the less educated before we can say that one of these variables causes the other.

This association criterion of causality is very much a statistical concern, and we have already dealt with it extensively. The bivariate techniques of Chapters 5 through 10 were directed at this causal condition. Bivariate cross-tabulation, comparison of means, analysis of variance, and regression and correlation are procedures to determine if variables X and Y are associated with each other. We can decide if a dependent variable is associated with an independent variable by comparing percentages or means across independent variable categories or by assessing scatterplots and regression and correlation coefficients.

That leaves the third condition of causality: The association between variables X and Y must not be due to some third, antecedent variable. An *antecedent variable* is one that occurs prior to the independent variable (and thus also prior to the dependent variable) in a causal chain. If there is such a prior variable that explains away the association between X and Y, then X and Y are not themselves causally connected. This third condition too is very much a statistical concern. This chapter offers some extensions of cross-tabulation that help us determine rationally whether or not this third condition is met. The next chapter offers an extension of regression and correlation analysis to address the same problem when both variables are measured at the interval/ratio level.

---

[1] Actually, there are some advanced statistical techniques that help specify causal ordering in ambiguous situations, but they are way beyond what we can do in this introductory text.

11.2   Spurious Relationships

Let's see how to use cross-tabulation with antecedent variables. In the workbook we will analyze real data, but it is best to begin with an imaginary example. Consider Table 11.1's relationship between the prevalence of storks and birth rates in 200 districts of some imaginary European country.

Table 11.1. Birth Rate by Number of Storks (in percentages)

Birth Rate	Number of Storks	
	Few	Many
High	44	62
Low	56	38
Total	100	100
(N)	(100)	(100)

$\chi^2 = 6.50$; $p < .05$; $G = .35$

Sure the example here is substantively ridiculous. I have deliberately chosen it for that reason. I know that storks don't bring babies. But a make-believe example frees me to make up whatever frequencies and percentages I want in order to illustrate the logic of control variables. Hypothetical examples are wonderfully free of the messy ambiguity we find in the real world. We have time enough for "reality" later . . . starting with the real data in the workbook.

So let's see what Table 11.1 tells us. It shows a clear relationship between the prevalence of storks and birth rates. Only 44 percent of areas with few storks have high birth rates, compared with 62 percent of areas with many storks—a moderate difference of 18 percentage points. Certainly the presence of storks is associated with high birth rates. Gamma is .35.[2] Despite a fairly small N, the chi-square test indicates that the relationship is significant at the .05 level.

The make-believe character of this relationship surely does not inhibit the more critical among us from objecting, "Wait a minute. Storks don't bring babies. The relationship isn't really causal. What's probably happening is that rural areas have a lot of storks and also have high birth rates, while towns have few storks and also low birth rates. I'll bet if we take into account location—whether areas are rural

---

[2] Gamma is equivalent to another measure of association, Q, for 2 by 2 tables. Q is usually used with nominal variables, however, so I am using gamma here and throughout most of this chapter.

or urban—then this apparent relationship between storks and birth rates disappears."

Well argued, young researcher. Maybe you're right. Let's find out by looking at the relationship between storks and birth rates separately for rural and urban areas. That way we can hold location constant while examining the stork–birth rate relationship. Maybe we would find the patterns shown in Table 11.2.

Table 11.2. Birth Rate by Number of Storks,
Controlling for Location (in percentages)

	Location			
	Rural		Urban	
	Number of Storks		Number of Storks	
Birth Rate	Few	Many	Few	Many
High	80	80	20	20
Low	20	20	80	80
Total	100	100	100	100
(N)	(40)	(70)	(60)	(30)
	$\chi^2 = 0.00$; n.s.		$\chi^2 = 0.00$; n.s.	
	$G = 0$		$G = 0$	

If so, you are right. Among rural areas, 80 percent with few storks and 80 percent with many storks have high birth rates. No difference there—among rural areas birth rate is not related to prevalence of storks. And there is no stork–birth rate relationship in urban areas either: For the cities, 20 percent with few storks have high birth rates and 20 percent with many storks have high birth rates. No difference—no relationship.

Notice that each half of the above table holds location constant in a nearly literal sense. All 110 (i.e., 40 + 70) cases on the left are rural; and all 90 (i.e., 60 + 30) cases on the right are urban. In effect, Table 11.2 consists of two bivariate tables, each showing the relationship between storks and birth rates for cases with a particular location. One bivariate table is for rural areas; the other is for urban. Whatever happens within each half of the table cannot be due to location since all the cases in that half have the same location. These bivariate tables embedded in a multivariate table are called *partial tables* or *conditional tables*. I'll use the former, more common term.

We can compute a measure of association for each partial table and interpret it just as we did for bivariate tables. With two ordinal variables, I have calculated gamma. (Other ordinal measures—D_{YX} or tau-b—would serve as well, depending on how we want to treat ties. Here I am choosing to ignore ties.) In Section 11.10 we will learn a measure of association called partial gamma that provides an overall measure of the strength of the independent variable–dependent variable relationship, controlling for a third variable. For now, however, we will rely only on the separate measures of association for each partial table.

We can also carry out a chi-square test for the overall independent variable–dependent variable relationship after controlling for a third variable, although here the statistics quickly become somewhat complex and a little beyond the reach of this text. In Table 11.2 I report two chi-square tests—one for each partial table. These tests work as expedient assessments of the statistical significance of partial relationships. We could also sum these chi squares and their degrees of freedom as a significance test for the overall relationship between the independent and dependent variables, controlling for the third variable.

But there are more sophisticated—and, frankly, far better—methods of computing chi square for partial tables that involve "partitioning" the bivariate chi square into separate chi squares for partial tables. These methods are analogous to the way in which analysis of variance breaks the total variance into the between-groups and within-groups variance. I refer you to more advanced statistics texts for these ways of handling chi-square tests for partial tables. Meanwhile, we will rely on "regular" chi-square tests for partial tables.

These tables suggest that the causal connections among variables look like those in Figure 11.1. We have seen in Table 11.1 that number of storks is related to birth rates—the apparent relationship on the left of Figure 11.1. However, the causal diagram on the right—what is really happening—describes location affecting both storks and birth rates, with no causal connection between the latter two variables.

Figure 11.1. Apparent and Real Relationships

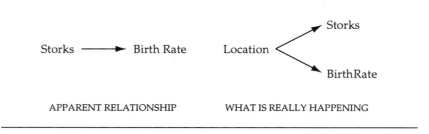

APPARENT RELATIONSHIP   WHAT IS REALLY HAPPENING

We can check the relationships of location to storks and of location to birth rates with bivariate Tables 11.3 and 11.4. Yes, just as we suspected: Rural areas have more storks than urban areas do (Table 11.3), and rural areas also have higher birth rates than urban areas (Table 11.4). Each relationship is strong, with gammas of —.56 and —.88 (treating location as ordinal, with rural as low and urban as high), and significant at the .001 level.

Table 11.3. Number of Storks by Location (in percentages)		
Number of Storks	Location	
	Rural	Urban
Many	64	33
Few	36	67
Total	100	100
(N)	(110)	(90)

$\chi^2 = 18.18$; $p < .001$; $G = -.56$

Table 11.4. Birth Rate by Location (in percentages)		
Birth Rate	Location	
	Rural	Urban
High	80	20
Low	20	80
Total	100	100
(N)	(110)	(90)

$\chi^2 = 71.54$; $p < .001$; $G = -.88$

11.3   Some Terminology

Some terminology is useful here to describe succinctly the logic of multivariate analysis. A variable that we hold constant while examining a bivariate relationship is called a *control variable* or *test factor*. Location (rural-urban) is a control variable in Table 11.2. If we find that a bivariate relationship largely disappears when we control for an antecedent variable, we say that the bivariate relationship is *spurious*. The stork–birth rate relationship shown in Table 11.1 is spurious.

We use the term *explanation* to describe the result or conclusion from this analysis. Explanation is a good term since that's exactly what we do: We *explain* the bivariate relationship by identifying the antecedent variable responsible for it. Once we control for location by holding it constant, as we do in each half of Table 11.2, the relationship between storks and birth rate disappears. It is a spurious relationship. We have explained it with the antecedent variable location.

The general process of introducing control variables is called *elaboration*. Explanation, then, is one result of elaboration—the result that reveals a bivariate relationship to be spurious. We will soon learn some other possible results of elaboration. When control variables are introduced with tables, as in this chapter, we call the process *table elaboration*. Paul Lazarsfeld, a famous sociologist who taught at

Columbia University, formalized the so-called elaboration model, and we owe much of our tabular terminology to him.

Two more terms: We refer to bivariate relationships as *zero-order* to distinguish them from relationships with control variables. The zero in zero-order means there are no (i.e., zero) control variables. For example, Table 11.1 is a zero-order table; so are Tables 11.3 and 11.4. There are no controls in any of those tables—they are just bivariate tables. When we introduce a control variable, the table for each category of the control variable is called a *partial table*. When we have only one control variable, as in Table 11.2, we call each subtable a *first-order partial table*. The rural half and the urban half of Table 11.2 are first-order partial tables. If we introduce two control variables simultaneously, we create *second-order partial tables*. And so on for as many control variables as we simultaneously introduce.

In these terms, explanation occurs when a relationship found in a zero-order table completely or largely disappears in partial tables. Then the original relationship is demonstrated to be spurious. We have explained the relationship. We do not insist, of course, that the relationship *completely* disappear in the partial tables since that is unlikely. We say a relationship is spurious as long as the relationships in the partial tables are very weak.

11.4 Examples of Spurious Relationships

Statistics and research methods texts contain many "classic" examples of spurious relationships. I can't resist sharing them with you. In fact, I have already mentioned some when I warned in Section 5.8 that association does not imply causation. These examples are about as silly as the spurious stork–birth rate example (which is one of these classics), but they help make the point that spurious relationships disappear when an antecedent control variable is introduced.

Here's the first example: The flooding of the Ganges River in India is related to the amount of street crime in New York City. When the Ganges spills over its banks, street crime goes up in New York. So we could reduce street crime in New York by flood control in India? Not a chance. Flooding doesn't cause American crime. It is just that the warmer part of the year brings both flooding to India and increased criminal activity to New York. Season is the antecedent variable explaining the flooding-crime relationship.

Another example: Towns whose priests, ministers, and rabbis earn high salaries tend to have greater liquor consumption than towns with poorly paid clergy. Well-paid clerics using their higher salaries to drink lots of booze? No, it's just that clergy salaries and liquor consumption are both related to the affluence of a community. Residents of wealthier towns can afford both to pay their clergy more and also to buy more alcohol. There is no causal connection between clergy salaries and liquor consumption. It's a spurious association.

Still another example: The more firefighters fighting a fire, the greater the property loss. Do the firefighters cause property damage? Firefighters fanning flames, fiendishly fostering fiscal failure? Nope. An antecedent variable—size of the fire—produces this spurious association. The bigger the fire, the more firefighters. The bigger the fire, the greater the property damage. Firefighters and property damage are not causally related—only spuriously associated.

In Section 5.8 I mentioned that children's spelling test scores are associated with their shoe sizes. You figure it out.

11.5   Replication

There is nothing in the elaboration process that ensures that a bivariate relationship is spurious. Maybe the bivariate relationship is "real." Maybe the independent variable actually does cause the dependent variable. Maybe, just maybe, storks really do bring babies.

If the stork–birth rate relationship in Table 11.1 is genuinely causal, a control for location produces first-order partial tables that might look like those in Table 11.5. Each of these partial tables closely replicates the bivariate table. Among rural areas, high birth rates are found in only 45 percent of areas with few storks but in 61 percent of areas with many storks. That is almost exactly what we found in our original bivariate table, so among rural areas storks are related to birth rates. I will spare you the specifics for urban areas, but you can easily see the same pattern of association between storks and birth rates. The gammas are about the same as in the zero-order table—.32 and .39, compared with .35 in Table 11.1.

In this example, controlling for location makes almost no difference in the relationship between the independent and dependent variables. The association between storks and birth rates is much the same regardless of whether we look at rural or urban areas. In this situation we say that we have engaged in *replication* and have found our original bivariate relationship to be *genuine*. Replication and genuine are good terms for this situation since in our partial tables we have indeed replicated (i.e., duplicated) the original bivariate relationship and thus found it to be genuine.

Elaboration, then, can produce either explanation or replication (as well as some other outcomes that I will describe a bit later). The "mechanics"—the kinds of partial tables we construct—are identical for explanation and replication. What differs is what the patterns of data imply. If the partial tables show no relationship (or a very weak one) between the independent and dependent variables, we have explanation—the original relationship is spurious. If the partial tables show much the same patterns as in the zero-order independent variable–dependent variable table, we have replication—the original relationship is genuine.

Table 11.5. Birth Rate by Number of Storks, Controlling for Location (in percentages)

	Location			
	Rural		Urban	
	Number of Storks		Number of Storks	
Birth Rate	Few	Many	Few	Many
High	45	61	43	63
Low	55	39	57	37
Total	100	100	100	100
(N)	(40)	(70)	(60)	(30)

$\chi^2 = 2.78; p > .05$ $\chi^2 = 3.20; p > .05$
G = .32   G = .39

With replication, of course, it is possible that the original relationship really is spurious and we simply have failed to control for the right antecedent variable. There is always the possibility that if we control for some other antecedent variable, the partial tables will reveal the original relationship to be spurious. I will discuss this problem a little later when I compare the elaboration process with randomization in experiments.

11.6   Somewhere Between Explanation and Replication

Explanation and replication are two extremes that we approach only occasionally in most analyses of real data. That's in part why I chose an imaginary example. Made-up data allow much clearer examples than we usually find in the messy real world. More often we find that controlling for an antecedent variable reduces but does not completely eliminate an apparent relationship. Table 11.6 presents partial tables that illustrate this situation.

Storks and birth rates are still related within each partial table of Table 11.6, but the relationship is not as strong as in the original bivariate Table 11.1. In fact, the percentage point differences are only half or less those of the corresponding differences in Table 11.1. The gammas for the partial tables are roughly half as large—.19 and .16, compared with .35 in the zero-order table. Location explains some but certainly not all of the relationship between storks and birth rates. This is not to say that storks really do cause babies, but only that location does not explain the entire relationship. Perhaps other antecedent variables (maybe age of homes or maybe size of homes) explain the rest. Or maybe the rest of the stork–birth rate relationship is genuine. Further analysis introducing additional control variables can assess these possibilities.

Table 11.6. Birth Rate by Number of Storks, Controlling for Location (in percentages)

	Location			
	Rural		Urban	
	Number of Storks		Number of Storks	
Birth Rate	Few	Many	Few	Many
High	48	57	47	53
Low	52	43	53	47
Total	100	100	100	100
(N)	(40)	(70)	(60)	(30)
	$\chi^2 = .95$; p = n.s.		$\chi^2 = .36$; p = n.s.	
	G = .19		G = .16	

This somewhere-between-explanation-and-replication is far more common than either pure explanation or pure replication. The reason is that the social world is a wonderfully complicated, complex network of multicausal connections. All sorts of variables are related to all sorts of other variables. Since we usually pick antecedent control variables that we have good reason to believe will "work," we are unlikely to introduce a control variable that has no effect at all on the original bivariate relationship—so pure replication is unusual. And likewise, the social world is so multicausal that an antecedent control variable is unlikely to explain all of the association between two other variables. After all, few dependent variables are related to only one prior independent variable, so pure explanation is unusual too.

11.7   Specification

Controlling for an antecedent variable may also reveal that the bivariate relationship is found for some value(s) of the control variable but not for others. Consider, for example, Table 11.7. The positive zero-order relationship between storks and birth rate is found among rural areas—in fact, it is somewhat stronger in rural areas than in the zero-order table (G = .47 versus G = .35). However, the relationship virtually disappears among urban areas (G = .03). We have *specified* where the stork–birth rate relationship occurs: It occurs among rural areas but not cities.

Table 11.7. Birth Rate by Number of Storks, Controlling for Location (in percentages)

	Location			
	Rural		Urban	
	Number of Storks		Number of Storks	
Birth Rate	Few	Many	Few	Many
High	48	71	42	40
Low	52	29	58	60
Total	100	100	100	100
(N)	(40)	(70)	(60)	(30)

$\chi^2 = 6.23$; p < .05 $\chi^2 = .02$; n.s.
G = .47   G = .03

For good reason, then, we call this elaboration process *specification*. Specification is a form of interaction. The relationship between the independent and dependent variables itself varies depending on the value of the control variable. Specification too is a fairly common outcome of the elaboration process. Many relationships are found to vary, for example, between men and women, or between African Americans and Whites, or among people living in different regions . . . and on and on.

11.8   Suppressor Variables

Explanation, replication, something in between, and specification—those processes do not exhaust reasons for introducing control variables or what we may find when we introduce them. The elaboration processes that we have considered all assume that a zero-order relationship exists. Our job then is to try to find an antecedent variable (or maybe more than one) that can explain or weaken or specify that original relationship.

But controls for antecedent variables are also appropriate for zero-order relationships that show *no* association between the independent and dependent variables. An antecedent variable may suppress the appearance of a relationship so that it does not show itself in a bivariate table. If we control for that antecedent variable, the partial tables may reveal the "true" relationship. Such an antecedent variable is called, appropriately, a *suppressor variable*.

Here's a grotesque example. In *real* fairy tales (not the nicey-nice Disney versions), lots of children get eaten by witches and trolls. Ever wonder if eating children gives indigestion to fairy-tale ghouls? Seems like it would. I can imagine witches around a caldron or trolls under a bridge, belching and complaining of upset tummies after devouring several particularly plump children who didn't mind their parents (or college students who didn't do their statistics exercises).

Let's put the food preference–indigestion relationship to the test. Table 11.8 is a zero-order table for the relationship between food preference and indigestion among 500 fairy-tale ghouls. Nope—no relationship between food preference and indigestion. Ghouls that do not prefer children are just as likely to have upset tummies as ghouls who prefer to eat kids. I guess eating kids does not cause upset stomachs for ghouls.

Table 11.8. Frequency of Indigestion by Food Preference (in percentages)

Indigestion Frequency	Food Preference	
	Does Not Prefer Children	Prefers Children
High	50	50
Low	50	50
Total	100	100
(N)	(300)	(200)

$\chi^2 = .00$; n.s.; lambda $= .00$

But wait a minute! There are two kinds of ghouls: witches and trolls. I bet witches are more likely than trolls to eat children (remember Hansel and Gretel's close escape). But I bet, too, that witches are less likely than trolls to get indigestion. If so, type of ghoul is a variable that suppresses the real relationship between eating children and indigestion.

Table 11.9 is a partial table testing this hunch. Yes, indeed, there is a relationship between food preference and indigestion, although the relationship is the reverse for witches and trolls. Among trolls, preference for children is associated with indigestion: 80 percent of trolls who savor children suffer high indigestion, compared with only 30 percent of trolls who don't like eating children. The opposite pattern appears among witches, with high indigestion plaguing only 20 percent of witches who like eating kids and a whopping 90 percent who do not eat little whippersnappers. What appears to be no association in the zero-order table appears to be a complex relationship when the control variable is introduced. Type of ghoul in this example acts as a suppressor variable masking the complex zero-order relationship between food preference and indigestion.

Table 11.9. Frequency of Indigestion by Food Preference, Controlling for Type of Ghoul (in percentages)

	Type of Ghoul			
	Troll		Witch	
	Food Preference		Food Preference	
Indigestion Frequency	Does Not Prefer Children	Prefers Children	Does Not Prefer Children	Prefers Children
High	30	80	90	20
Low	70	20	10	80
Total	100	100	100	100
(N)	(200)	(100)	(100)	(100)

$\chi^2 = 66.96$; $p < .001$
lambda = .43

$\chi^2 = 98.99$; $p < .001$
lambda = $-.67$

Incidentally, although I won't do so here, you can construct two additional zero-order tables from the marginal frequencies of Table 11.9: food preference by type of ghoul and frequency of indigestion by type of ghoul. You will find that these bivariate tables show that witches are more likely than trolls to prefer eating children and less likely to suffer indigestion.

The moral of this section: If a bivariate relationship fails to appear in a zero-order table when you expect it to, consider possible suppressor variables that may be at work.

11.9   Controlling for an Intervening Variable

So far we have controlled for antecedent variables. But we also can use the same elaboration "mechanics" to control for an *intervening variable* that we regard as causally linking an independent to a dependent variable. The second, or temporal, condition of a causal relationship requires that the intervening variable occur after the independent variable and before the dependent variable in a causal chain (see Section 11.1). Schematically, a causal chain with an intervening variable looks like this:

Independent ————————→ Intervening ————————→ Dependent
Variable   Variable   Variable

If we control for a variable that we believe links an independent and a dependent variable, three things may happen: The original relationship may disappear in the partial tables; the original relationship may continue to be found within the partial tables; or something in between may happen. Let's consider each of these situations in turn.

First, if the original relationship disappears within the partial tables controlling for a variable we believe is intervening, we conclude that the intervening variable links the independent and dependent variables. The patterns in the partial tables resemble those we found for spurious relationships, although for quite different causal reasons. And just as we concluded for spurious relationships that the antecedent variable explains the entire relationship between the independent and dependent variables, now we conclude that the intervening variable entirely accounts for how the independent variable affects the dependent variable.

This pattern does not mean that the independent variable and dependent variable are not related. To the contrary, they certainly are related. We have now demonstrated how the independent variable causes the dependent variable. It does so via the intervening variable. The technical name for this situation is *interpretation*. We say that the intervening variable *interprets* the relationship between the independent and dependent variables.

But, on the other hand, what if the partial tables look much like the original table? Then we conclude that the alleged intervening variable really does not intervene. It does not link the independent and dependent variables. The control variable just doesn't "work" as part of the causal chain. This is not to conclude that there are no intervening variables but only that we have not successfully identified one.

Finally, a third possibility. As with antecedent variables, real situations controlling for intervening variables often lie somewhere between these two extremes of completely interpreting a causal relationship and not interpreting the relationship at all. We often find that a plausible intervening variable interprets some but not all of the relationship between an independent variable and a dependent variable. The intervening variable matters, but there must be some other variables that also causally connect the independent and dependent variables. Yes, the world is an intricate web of causal interconnections.

If we do find a variable linking an independent and dependent variable, our job is not done. We can then seek additional variables interpreting the links between the independent and intervening variables and between the intervening and dependent variables. And so on and so on as we fill in the causal web.[3] Frankly, there are far more efficient ways than cross-tabulation to analyze causal relationships, but the logic underlying more advanced methods is much the same as the logic of table elaboration.

I want to call special attention to the importance of theoretical models that we bring to statistical analyses. The logic and procedures of elaboration make no distinction between the following two models:

Model A   Model B

Data analyses relating variables X and Y while controlling for Z yield the same results regardless of whether the spuriousness of Model A or the causal chain of Model B describes the real relationship among variables. Data analyses will not lead us to choose between Model A and Model B. That choice needs to be based on our paradigms, our theories, and sometimes even our common sense.

11.10   Partial Gamma

We found gamma a useful measure of association for cross-tabulations involving ordinal or categorized interval/ratio variables. *Partial*

---

[3] Causal chains raise fascinating philosophical and applied issues as they lengthen. Does the flapping of a butterfly's wings in China affect weather in Michigan? Or remember the children's verse: "For loss of a nail a shoe was lost/For loss of a shoe a horse was lost/For loss of a horse a rider was lost/For loss of a rider a battle was lost/For loss of a battle a war was lost/For loss of a war a nation was lost." Do causal chains go on forever or do they eventually "peter out"? At what point, if any, do we say that one event is just too far removed from another, just too distant in a causal chain, to be a cause? Much debate over the impact of historical events on contemporary social problems involves just such issues.

gamma (symbolized G_p) is an extension of gamma to multivariate tables. Partial gamma describes the strength of the relationship between an independent variable and dependent variable measured at the ordinal level after the effects of a control variable have been removed. Like "regular" zero-order gamma, partial gamma is a PRE measure of association. It reports the proportion that we can reduce errors in predicting the ordering of pairs of scores if we know the independent variable scores, after controlling for the effects of the control variable.

Partial gamma's calculation is a straightforward generalization of gamma. Recall the formula for gamma in Section 7.5:

$$G = \frac{\text{Same} - \text{Opposite}}{\text{Same} + \text{Opposite}}$$

where Same and Opposite are the number of pairs of cases ordered on the independent and dependent variables in the same and in the opposite direction. For partial gamma, we count the same and opposite pairs in each partial table following exactly the same procedure we used for bivariate tables (see Section 7.5). We then sum the number of same pairs for all partial tables; and likewise, we sum all opposite pairs for all partial tables. Finally, we compute partial gamma with this formula:

$$G_p = \frac{\Sigma \text{Same} - \Sigma \text{Opposite}}{\Sigma \text{Same} + \Sigma \text{Opposite}}$$

The summation sign Σ here means that we sum pairs across all partial tables.

As an example, let's find partial gamma for the first-order partial relationship described in Table 11.6. This table describes the relationship of birth rate to number of storks, controlling for location. Here are the frequencies underlying the percentages in Table 11.6:

	Location			
	Rural		Urban	
	Number of Storks		Number of Storks	
Birth Rate	Few	Many	Few	Many
High	19	40	28	16
Low	21	30	32	14

We calculate gamma by first finding the number of pairs of cases with variables ordered in the same direction and in the opposite direction:

$$\text{Same} = (21)(40) + (32)(16)$$
$$= 840 + 512$$
$$= 1352$$
$$\text{Opposite} = (19)(30) + (28)(14)$$
$$= 570 + 392$$
$$= 962$$

Then:

$$G_p = \frac{\Sigma\text{Same} - \Sigma\text{Opposite}}{\Sigma\text{Same} + \Sigma\text{Opposite}}$$

$$= \frac{1352 - 962}{1352 + 962}$$

$$= \frac{390}{2314}$$

$$= .168$$

$$= .17$$

This partial gamma of .17 indicates a fairly weak relationship between birth rate and number of storks, even after controlling for location. For pairs of cases, we reduce errors by 17 percent by "guessing" that the pair member from an area with many storks has the high birth rate, after controlling for location. As reported in Table 11.6, the gammas are .19 for the rural areas and .16 for urban areas. Partial gamma is a weighted average of these zero-order gammas. Note from this example that partial gamma is appropriate even if the control variable is nominal as long as the independent and dependent variables are measured at least at the ordinal level.

11.11   An Overview of Elaboration

It is helpful at this point to take stock of what may happen when we carry out a multivariate cross-tabulation of the relationship between an independent variable (IV) and a dependent variable (DV), controlling for a third variable (CV). Table 11.10 offers a summary of situations we encounter in table elaboration.

Table 11.10. Summary of Table Elaboration

Control Variable	Zero-Order IV-DV Relationship?	Partial Tables Controlling for IV	Conclusion
Antecedent	Yes	No relationship in partial tables. $G_p \approx 0$	IV-DV relationship is spurious. (Explanation)
Antecedent	Yes	Similar to zero-order IV-DV table. $G_p \approx G$	IV-DV relationship is genuine. (Replication)
Antecedent	Yes	Weaker than zero-order IV-DV table. $0 < G_p < G$	CV explains part but not all of IV-DV relationship.
Antecedent	Yes	Varies from one partial table to another. Gs vary among partial tables.	IV-DV relationship varies depending on value of CV. (Specification)
Antecedent	No	Relationship in partial tables. $G_p > G \approx 0$	CV is a suppressor variable masking IV-DV relationship.
Intervening	Yes	No relationship in partial tables. $G_p \approx 0$	CV causally links the IV and DV. (Interpretation)
Intervening	Yes	Similar to zero-order IV-DV table. $G_p \approx G$	CV does not link IV and DV.
Intervening	Yes	Weaker than zero-order IV-DV table. $0 < G_p < G$	CV partially links IV and DV, but there are also other linking variables.

11.12 Elaboration and Problems of Small Ns

Table elaboration is inefficient in that it requires a large number of cases. The more control variables and the more values of the independent and control variables, the more cases needed. Without sufficient cases, percentages in partial tables will be based on too few cases for us to have much confidence in our findings. Expected frequencies may also be too small (less than 5) to carry out chi-square tests. Even in the very unlikely situation in which cases are evenly divided between partial tables, percentages based on 100 cases in a zero-order table become based on 50 cases in a partial table controlling for a dichotomous variable. And if a second dichotomous variable is introduced, those 50 cases are reduced to only 25 as the basis for percentages and so on. The problem is even more severe for nondichotomous control variables (i.e., control variables with more than two possible values). Yes, table elaboration quickly spreads all but the largest Ns very thin.

We can ameliorate (although not eliminate) problems of small and diminishing Ns by collapsing independent and control variables. We may also reduce small N problems by excluding independent or control variable categories that have few cases. We have seen these data management strategies before with bivariate tables, and they work as well with multivariate analyses. As in bivariate analyses, of course, we shouldn't collapse or exclude categories if doing so obscures important detail or makes no theoretical sense.

Ultimately, however, no data management techniques can solve problems raised by table elaboration's inefficiency. As a practical matter, we are usually restricted to introducing only one or maybe two control variables unless our initial N is very, very large. Still, within these limitations, table elaboration offers a useful, sometimes even essential, set of procedures for handling antecedent and intervening variables.

11.13   The Relationship of Multivariate Analysis to Experimental Design

In true experiments, a researcher uses randomization to assign subjects to experimental and control groups entirely on the basis of chance. Within the limits of chance, random assignment of subjects ensures that experimental and control groups are identical (or very close to it) with regard to every possible antecedent variable: gender, race, ethnicity, and so on . . . including even silly variables that are unlikely to matter such as favorite cartoon character or veggie preference . . . or whatever. Experimental and control groups are even identical (or nearly so) on variables that you have never heard of and will never hear of. (Sorry, but I can't offer you any examples of variables you will never hear of.) Experimental and control groups are thus equalized (or close to it) with respect to any and all variables that could possibly affect the relationship between independent and dependent variables. Chance "works" better with larger numbers, so the larger the number of subjects in an experiment, the closer the researcher is likely to get to perfectly equating experimental and control groups.

Randomization is the major advantage of experiments: Their very designs implicitly impose controls for all possible antecedent variables. After the independent variable is introduced, whatever dependent variable differences a researcher then finds between experimental and control groups cannot be due to any antecedent variable. Within the limits of chance, randomization equates the experimental and control groups on all antecedent variables.

But sometimes we can't do experiments. For practical or ethical reasons, we may not be able to randomly assign subjects to experimental and control groups. If we are studying the effects of gender, for example, we cannot randomly assign some subjects to be male and some to be female. (Any volunteers for this experiment?) If we are studying the effects of education, we cannot randomly assign some subjects to go to college and other subjects to stop their educations after high school. (Again, any volunteers?) Instead, we have to take people as they are and do the best we can using nonexperimental research designs like surveys.

That is where multivariate analysis techniques come in. The use of control variables approximates randomization in experimental designs. Introduction of an antecedent control variable eliminates the effects of that variable on the relationship between the independent and dependent variables. This is exactly what randomization does too, although randomization does so much more efficiently by simultaneously eliminating the effects of *all* antecedent variables. Multivariate analysis proceeds much more slowly, controlling for the effects of antecedent variables one or, at most, a few at a time.

This means that multivariate analysis can never give us as much confidence that a relationship is genuine as true experimental designs can. Even if a relationship "holds" after we control for an antecedent variable, there is always the possibility that some other, uncontrolled antecedent variable might explain the relationship. So we may then introduce a control for a second antecedent variable. And perhaps a third, or maybe even a couple more. However, data sets are always finite, so we eventually run out of antecedent variables to control for. And besides, we might not even know what the "right" antecedent variable is.

But we do the best we can. No sense whining or wringing our hands—there's too much research to be done. We try, via the theories and other ideas that inform our research, to include as many plausible antecedent variables as we can when we collect our data. We then control for these variables in our analyses. Certainly our confidence that a relationship is genuine increases with each control we introduce, but we have to accept that we cannot control for every possible variable except by using experimental designs. But, again, we do the best we can, and if our theories, our research designs, and our statistical analyses are solid, our best will be very good indeed.

11.14   Summing Up Chapter 11

Here is what we have learned in this chapter:

- An antecedent variable occurs causally prior to both the independent and dependent variables.

- Elaboration is the process of analyzing a bivariate relationship after removing the effects of one or more control variables. When percentage tables are used, this process is called table elaboration.

- A partial relationship removes the effects of a third variable, either antecedent or intervening.

- If a bivariate relationship is not found in partial tables that control for an antecedent variable, the bivariate relationship is spurious—it is explained by the antecedent variable. This process is called explanation.

- If a bivariate relationship is also found in partial tables that control for an antecedent variable, the bivariate relationship is genuine (pending the introduction of other antecedent control variables). This process is called replication.

- Often a control for an antecedent variable reduces but does not completely eliminate a relationship between two variables, indicating that the control variable explains part (but only part) of the relationship.

- If partial tables vary from one another, the control variable specifies the bivariate relationship. This process is called specification.

- An intervening variable occurs causally between an independent and dependent variable.

- If a control for an intervening variable largely eliminates a relationship, the control variable causally links the independent and dependent variables. This process is called interpretation.

- If a control for an intervening variable reduces but does not eliminate a relationship, the control variable causally links the independent and dependent variables, but there may also be other linking variables.

- Partial gamma is a PRE measure of association that summarizes the relationship between categorized ordinal or interval/ratio variables, controlling for the effects of a third variable.

- Controls for antecedent variables in multivariate analysis approximate (but cannot duplicate) the use of randomization in experimental research.

Key Concepts and Procedures

Ideas and Terms

multivariate cross-tabulation
partial table
causal relationship
antecedent variable
control variable *or* test factor
spurious relationship
explanation
elaboration and the elaboration model
table elaboration

zero-order relationship
first-order partial table
replication
genuine relationship
specification
suppressor variable
intervening variable
interpretation
partial gamma

Symbol

G_p

Formulas

$$G_p = \frac{\Sigma Same - \Sigma Opposite}{\Sigma Same + \Sigma Opposite}$$

ANALYSIS WRITE-UP 7: MULTIVARIATE CROSS-TABULATION

◆   ◆   ◆   ◆   ◆

We have previously seen (in Write-up 3) that support for affirmative action for women is much more strongly supported by Democrats and Independents than by Republicans. Table 1 describes this relationship. This difference between Republicans and Democrats in support for affirmative action may be due to a "gender gap" in support for political parties, with women tending Democratic and men tending Republican.

TABLE 1 ABOUT HERE

Table 2 introduces a control for gender. The pattern found in the original table is replicated in the partial tables. For both men and women, we again find similar patterns for Democrats and Independents, and Republicans are far less likely than either to support affirmative action for women. The relationship is somewhat stronger for males than females (respective gammas: .34 and .27). Thus, gender differences in political party preferences do not account for party differences in support for affirmative action for women.

TABLE 2 ABOUT HERE

Include the table at the end of the paper or report:

Table 1. Affirmative Action for Women by
Political Party Preference (in percentages)

Affirmative Action for Women	Political Party Preference		
	Democrat	Independent	Republican
Agree	64.2	61.4	42.0
Neither	12.7	15.5	15.1
Disagree	23.1	23.2	42.9
Total	100.0	100.1	100.0
(N)	(685)	(207)	(517)

$\chi^2 = 71.678$; df = 4; p < .001; G = .32

Table 2. Affirmative Action for Women by Political Party Preference,
Controlling for Gender (in percentages)

	Gender					
	Male			Female		
Affirmative Action	Political Preference			Political Preference		
for Women	Dem	Indep	Rep	Dem	Indep	Repub
Agree	60.2	57.0	36.3	66.7	64.9	48.0
Neither	12.7	16.1	15.4	12.7	14.9	14.8
Disagree	27.0	26.9	48.3	20.6	20.2	37.2
Total	99.9	100.0	100.0	100.0	100.0	100.0
(N)	(259)	(93)	(267)	(426)	(114)	(250)

$G_p = .30$ $\chi^2 = 36.398; df = 4;$ $\chi^2 = 25.812; df = 4;$
 $p < .001; G = .34$ $p < .001; G = .27$

CHAPTER 12
Multiple Regression and Correlation

In Chapter 10 we learned regression and correlation techniques to describe relationships between two variables measured at the interval/ratio level. This chapter expands on those bivariate techniques to include additional variables. Here we meet the multiple correlation coefficient, a statistic that measures the simultaneous effect of two or more independent variables on a dependent variable. We also encounter the beta coefficient, which measures the effect of an independent variable while taking into account the other independent variables. Along the way we will learn to test for the significance of multiple correlations. And, too, we will learn to create dummy variables to extend correlation analysis to nominal variables.

After this chapter you will be able to:

1. Understand the extension of regression analysis to two or more independent variables.

2. Recognize conditions under which multiple regression procedures are appropriate.

3. Calculate and interpret multiple correlation coefficients.

4. Calculate and interpret beta coefficients.

5. Carry out and interpret significance tests for multiple correlations.

6. Create and explain dummy variables.

12.1 Extending the Regression Model

For the 50 most populous nations, correlations reported in Chapter 10 (Section 10.12) described fairly strong negative relationships between fertility rate and both urbanism and radios per 100 persons. The percent of population living in urban areas explains 31 percent of the variation in fertility rates (i.e., $r^2 = .31$) and radios per 100 persons explains about 22 percent of the variation in fertility. But how much variation in fertility is explained by both urbanism and radio availability taken together? We cannot simply add the percentages of variation explained by urbanism and radio availability individually because these independent variables are themselves correlated ($r = .62$). Their effects on fertility thus overlap, and we need some way to take that overlap into account in measuring the combined effect of urbanism and radio availability. We also need a way to measure the effect of each independent variable on the dependent variable while controlling for the other independent variable.

We can measure the effects of two or more independent variables on a dependent variable by extending the linear regression model presented in Chapter 10. Let's plot data points for our three variables in a three-dimensional scatterplot, with independent variables (percent urban and radios per 100 persons) arrayed along two axes and the dependent variable (fertility rate) on the third axis. As with bivariate scatterplots, it is conventional to locate the dependent variable on the Y-axis. This scatterplot is fairly easy to visualize. Think of it as the corner of a room. The floor is the scatterplot for the urbanism-radios relationship. One wall is the scatterplot for the urbanism-fertility relationship. The other wall plots the radios-fertility relationship. We plot each case in the three-dimensional space at the intersection of the case's scores on the three variables.

Take Egypt as an example. Data for the largest nations report that 45 percent of Egyptians live in urban areas, there are 25 radios for every 100 Egyptians, and Egyptian women average 4.35 children. Therefore, we locate Egypt's scatterplot dot 45 units on the urbanism axis, 25 units on the axis for radio availability, and 4.35 units up on the fertility axis. Figure 12.1 shows Egypt's position in this three-dimensional scatterplot that, as I suggested above, you can visualize as the corner of a room. Now imagine all 50 data points similarly located in this three-dimensional space. That would be a cluttered graph with all 50 cases, so I won't draw it here. But our minds work more clearly than 2-D paper when it comes to imaging 3-D graphs, so I think you can easily imagine such a scatterplot for 49 most populous nations. That's 49 rather than 50 because we have to exclude one country —Uzbekistan—that is missing data on radios per 100 persons. Deletion will be listwise, with Uzbekistan consistently deleted throughout the analysis.

Figure 12.1. Three-Dimensional Scatterplot (Egypt only)

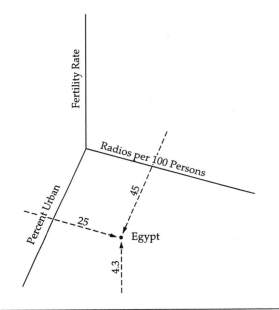

What we need is an efficient way to summarize relationships among variables shown in the scatterplot. Think back to the bivariate regression model of Sections 10.1 and 10.2. We summarized a bivariate relationship with a regression line that has this equation:

$$Y = a + bX$$

where Y = score on the dependent variable

   a = Y-intercept or constant (the point at which the regression line crosses the Y-axis)

   b = slope, or the change in Y for every one-unit increase in X

   X = score on the independent variable

This regression line is the best fit of the scatterplot of scores in the sense that it minimizes the sum of the squared deviations between actual dependent variable scores and scores predicted by the regression line.

   Now let's extend the notion of a regression line to a dependent variable regressed onto *two* independent variables. Just as we passed a best-fitting line through a two-dimensional scatterplot, so we can pass a flat two-dimensional plane through three-dimensional space. The plane cuts through the data points in such a way as to minimize the sum of the squared distances (in the vertical or dependent variable direction) between each data point and the plane. Schematically this least-squares plane looks something like the plane passing through the three-dimensional scatterplot in Figure 12.2.

Figure 12.2 Regression Plane in Three-Dimensional Scatterplot

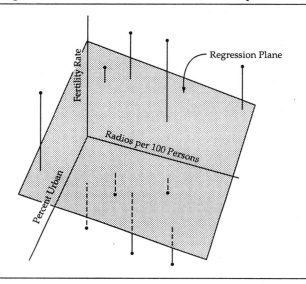

I have plotted ten data points in Figure 12.2, five below and five above the regression plane. I have also drawn vertical lines showing the distance between each data point and the regression plane. These distances are residuals, or the differences between fertility rate scores and scores predicted by the regression plane. We saw residuals on a two-dimensional graph of a bivariate relationship. Here they are in a three-dimensional graph of a multivariate relationship. As in bivariate regression, these residuals are errors—departures of actual scores from scores predicted by the best-fitting multivariate plane.

Just as we represent a bivariate regression line with the equation $Y = a + bX$, we can represent a multivariate plane with this *multiple regression equation*:

$$Y = a + b_1X_1 + b_2X_2$$

where Y = score on variable Y

  a = Y-intercept or constant (the point at which the regression plane crosses the dependent variable axis)

  b_1 = slope of the plane with respect to independent variable X_1

  X_1 = score on independent variable X_1

b_2 = slope of the plane with respect to independent variable X_2

X_2 = score on independent variable X_2

In our example, Y represents the fertility rate (the dependent variable) and X_1 and X_2 stand for percent urban and radios per 100 persons, respectively. This regression equation predicts fertility rate on the basis of what we know about a country's urbanism and radio availability. The b_1 and b_2 coefficients are *unstandardized regression coefficients*. The adjective "unstandardized" distinguishes these coefficients from standardized ones that we'll encounter later in this chapter, although researchers often call them simply regression coefficients if the context makes clear that these are the kind of regression coefficients referred to. Unstandardized regression coefficients are partial slopes that describe the change in the dependent variable Y associated with an increase of one unit in the independent variable, controlling for the other independent variable(s).

We calculate the intercept a (sometimes called the constant) and the unstandardized regression coefficient b_i from values of zero-order regression coefficients, but the formulas are a little complicated and we won't bother with them. We'll just use the computer to find them. In fact, I have done that and found this equation describing the regression of fertility rate on percent urban and radios per 100 persons.

$$Y' = 5.59 - .032X_1 - .010X_2$$

where Y' = fertility rate

X_1 = percent urban

X_2 = radios per 100 persons

The negative signs on the regression coefficients indicate that urbanism and radio availability have negative effects on fertility. The more urbanism and the more radios, the lower the fertility rate. The magnitudes of the unstandardized regression coefficients describe the size of the effect of each variable on fertility rate, controlling for the other independent variable. Thus, with radios per 100 persons controlled, a one percentage point increase in urbanism decreases fertility rate by .032 child per woman. Similarly, with urbanism controlled, an increase of 1 radio per 100 persons is associated with a reduction in the fertility rate of .010 child per woman.

The regression equation can be used to predict dependent variable scores—fertility rates in our example. We can, for example, plug Egypt's urbanism and radio availability scores into the multiple regression equation:

$$Y' = 5.59 - .032X_1 - .010X_2$$
$$= 5.59 - .032(45) - .010(25)$$
$$= 5.59 - 1.44 - .25$$
$$= 3.90$$

As in bivariate regression, the prime on Y' reminds us that we are predicting the fertility rate. Our knowledge of Egypt's urbanism and radio availability leads us to predict a fertility rate of 3.90 for Egypt. Egypt, in fact, has a fertility rate of 4.35. So Egypt's fertility rate is .45 higher than we would expect on the basis of its urbanism and radios per 100 persons (4.35 − 3.90 = .45). Our regression equation is off just this small amount in estimating Egypt's fertility rate. But we have nevertheless improved our estimate of Egypt by using this information about independent variables. Without knowledge of percent urban and radios per 100 persons, our "best guess" of Egypt's fertility rate would be 3.56, the grand mean for the 49 countries. Egypt's actual fertility rate is .79 higher than that crude prediction based on the mean (4.35 − 3.56 = .79).

There are no new statistical ideas here—just the extension of old ideas from bivariate regression. However, there are some assumptions made in this model beyond the interval/ratio level of measurement that is applied to all regression approaches considered in this text. Just as in the bivariate situation, we assume that the independent variables are linearly related to the dependent variable. After all, a flat plane is an extension of a one-dimensional straight line over two dimensions. If the relationships of the independent variables to the dependent variable are not linear, this model may poorly describe the actual relationship. A flat regression plane, for example, cannot describe a saddle-shaped curvilinear relationship any better than a straight line can describe a U-shaped bivariate relationship. Summary measures based on a linear model will misrepresent such a relationship by underestimating its strength. As with bivariate regression, curvilinear relationships may be "straightened out" by transforming variables, but these techniques are a little beyond this text.

A second assumption underlying the multiple regression model is less obvious: The effects of the independent variables on the dependent variable are additive, with no *statistical interaction* between them. We assume, for example, that although both urbanism and radio availability are related to fertility, there is no combination of urbanism and radio availability that has an "extra" effect. Countries that are more urban and have more radios may have lower fertility rates, but we assume that the particular combination of both high urbanism and numerous radios does not have any additional, "extra" impact on lowering fertility.

A third assumption is also far from obvious: We assume that the independent variables in the model are uncorrelated with each other. Rarely is this assumption completely true (and it certainly isn't in this example), so it is fortunate that the multivariate regression model is robust enough to allow some correlations between independent variables without seriously compromising the model. But we must be careful when correlations between independent variables are very large—a situation statisticians refer to as *multicollinearity*. It is difficult, even impossible, to separate the effects of independent variables that are highly correlated. Obviously this assumption can be examined with bivariate correlational techniques.

In sum, we assume interval/ratio levels of measurement, linear relationships, additive independent variable effects (i.e., no interaction), and independent variables that are not strongly correlated. That's quite a set of assumptions, although we will add even more when we test for the statistical significance of a multiple correlation coefficient.

12.2   Multiple Correlation Coefficient

But significance tests are a couple sections away. We need first to learn about the *multiple correlation coefficient*, a measure of the combined effect of a set of independent variables on a dependent variable. We might use a multiple regression coefficient, for example, to gauge the combined effect of urbanism and radio availability on fertility rates in the world's largest nations. In other words, the multiple correlation coefficient, analogous to r for bivariate relationships, describes how closely data points are to the regression plane. The multiple regression coefficient is symbolized $R_{Y \cdot 12...}$, with Y denoting the dependent variable and the numerical subscripts that follow the • dot representing the independent variables $X_1, X_2 \ldots$. Subscripts are typically omitted if their identity is clear from the context.

Like all well-behaved measures of association, R equals 1.00 for a perfect relationship and is never greater than 1.00. Unlike a bivariate correlation coefficient, the multiple correlation coefficient is always equal to or greater than zero. A negative multiple correlation coefficient makes no sense since we cannot speak of the direction of a relationship involving more than one independent variable. Some independent variables may have positive relationships and some may have negative relationships with the dependent variable, but with multiple correlation we are interested only in the magnitude of their combined effects, not their directions. Therefore, R ranges from 0 to 1.00.

Analogous to other kinds of correlation coefficients, the square of the multiple correlation coefficient—$R^2_{Y \cdot 12...}$—reports the proportion of the variation in the dependent variable explained by the entire set of independent variables $X_1, X_2, \ldots$. $R^2_{Y \cdot 12\ldots}$ is sometimes called the *coefficient of multiple determination*, although mostly it is called simply **R-squared**. The proportion of variation in the dependent variable left unexplained by the set of independent variables is equal to $1 - R^2_{Y \cdot 12\ldots}$.

We learned in Chapter 4 to express variation as a sum of squares. The total variation in a dependent variable Y is measured as variation from the mean, and thus can be expressed as $\Sigma(Y_i - \overline{Y})^2$. The sum of squares explained by the regression of the dependent variable on a set of independent variables measures variation in predicted scores from the mean—easily expressed as $\Sigma(Y' - \overline{Y})^2$ where Y' stands for scores predicted by the multiple regression equation. We found that predicted score for Egypt—3.90. Now imagine predicting scores for all 49 countries, subtracting the grand mean from each, squaring the differences, and then adding them all up. That sum of squares would measure the variation in the dependent variable explained by the independent variables. The variation left unexplained—the residual —is based on differences between scores and estimated scores and is expressed by $\Sigma(Y_i - Y')^2$. Yes, this is still another sum of squares.

This should all sound pretty familiar. Note the direct analogy to the sums of squares used in the analysis of variance. In fact, we can most easily summarize these multiple regression sums of squares in a table analogous to an ANOVA table. Finding sums of squares is burdensome, and we won't trouble ourselves with their actual calculation. Fortunately, computers easily calculate these sums of squares for us. Here are the sums of squares for our example:

Source	Squares	df	Mean Sum of Squares	F	p
Regression	56.527	2	28.263	11.591	.001
Residual	112.167	46	2.438		
Total	168.694	48			

The regression sum of squares is the sum of squares explained by the set of independent variables. Since R^2 is the proportion of variance in the dependent variable explained by the independent variables, we calculate R^2 by expressing this regression sum of squares as a proportion of the total sum of squares, like this:

$$R^2 = \frac{\text{Regression Sum of Square}}{\text{Total Sum of Squares}}$$

$$= \frac{56.527}{168.694}$$

$$= .335$$

$$= .34$$

What about the df and mean sum of squares, F, and p? They are given in the preceding table because we'll need them a little later when we carry out a significance test for R^2.

12.3 Standardized Regression Coefficients (Beta Coefficients)

Regression equations describe the relationships of dependent variables to sets of independent variables, and those equations' unstandardized regression coefficients measure the effects of independent variables on dependent variables. However, the magnitudes of regression coefficients depend on variables' units of measurement. Radio availability, for example, is measured as radios per 100 persons. Obviously the regression coefficient would be different if the variable were based on radios per 10 persons or radios per 1000 persons or whatever.

There is nothing at all wrong with this—it's the nature of unstandardized regression coefficients—but it makes comparisons difficult when variables are based on different units of measurement. Consider: The means for percent urban and radios per 100 persons are 52.4 and 35.9, respectively. Their standard deviations are 25.1 and 39.0. How can we compare these numbers when they are based on different measurement scales? We're talking statistical apples and oranges here! Folk wisdom is right that we can't compare different fruit. Nor can we directly compare percentage points and radios per 100 persons.

Clearly we need a measure of the effects of independent variables that adjusts for different units of measurement while controlling for other independent variables. There is such a statistic, appropriately named the *standardized regression coefficient*, and often called a *beta coefficient* or a *beta-weight*. Betas are symbolized as $\beta_{Y1 \cdot 2}$ for the regression of Y on X_1, controlling for X_2. (For additional variables, we extend the symbol by tacking on subscripts for additional control variables after the • dot as needed.) The beta coefficient describes the effect of an independent variable on the dependent variable in standard deviation units. More specifically, the beta coefficient reports the standard deviation change in the dependent variable for a one standard deviation increase in the independent variable, controlling for the effects of other independent variables. Here are the formulas for beta coefficients with one control variable:

$$\beta_{Y1\cdot2} = \frac{r_{Y1} - r_{Y2}r_{12}}{1 - r_{12}^2} \quad \text{and} \quad \beta_{Y2\cdot1} = \frac{r_{Y2} - r_{Y1}r_{21}}{1 - r_{21}^2}$$

Formulas are more complicated for higher order betas—that is, betas with additional control variables. We will let computers work those out for us. But in our simple example with two independent variables:

$$\beta_{Y1\cdot2} = \frac{-.556 - (-.471)(.617)}{1 - (.617)^2} \quad \text{and} \quad \beta_{Y2\cdot1} = \frac{-.471 - (-.556)(.617)}{1 - (.617)^2}$$

$$= \frac{-.556 + .291}{1 - .381} \qquad\qquad\qquad = \frac{-.471 + .343}{1 - .381}$$

$$= \frac{-.265}{.619} \qquad\qquad\qquad\qquad = \frac{-.128}{.619}$$

$$= -.428 \qquad\qquad\qquad\qquad\qquad = -.207$$

Consider the beta coefficient for percent urban: −.428. This beta tells us that the dependent variable, fertility rate, decreases on average .428 standard deviation if the independent variable urbanism increases one standard deviation, controlling for radios per 100 persons. So, with radio availability held constant, an increase of 1 standard deviation in percent urban reduces fertility rate by .428 standard deviation. We can find out how much .428 standard deviation is in children per woman (which is how fertility rate is measured) by multiplying it by 1.875, the standard deviation of fertility rate: (.428)(1.875) = .80. The standard deviation of percent urban is 25.1. So, with radio availability held constant, an increase of about 25 percentage points in percent urban is associated with a decrease of about .80 children per woman.

Likewise, radio availability's beta coefficient of −.207 means that, with a control for percent urban, every standard deviation increase in radio availability, we expect a decrease of .207 standard deviation in fertility rate. I'll let you convert this beta into changes in number of children per woman. (Standard deviations of fertility rate and radios per 100 population are 1.875 and 38.997, respectively.)

By measuring the effects of variables in terms of standard deviations, betas provide a useful way to compare the relative effects of variables. We see in this example that urbanism has over twice the effect that radio availability has on fertility rates in the most populous countries.

Incidentally, a beta coefficient, unlike a correlation coefficient, can be greater than 1.00 since a one standard deviation change in an independent variable may produce more than one standard deviation change in the dependent variable.

12.4   Significance Tests for Multiple Correlation Coefficients

I have already listed several assumptions underlying the multiple regression model: interval/ratio-level variables; linear relationships; additive independent variables; and independent variables that are not correlated too highly. We can carry out significance tests for multiple correlation coefficients if we can make three additional assumptions. First, our data are from a random sample of the population. Second, dependent variable scores are distributed normally within values of the independent and control variables. And third, variances of the dependent variable are equal within values of the independent and control variables (i.e., homoscedasticity, which we first encountered in Section 9.2).

The significance test for R^2 (and thus R as well) uses the F distribution. As in analysis of variance, the value of F for R^2 is given by the ratio of the mean explained sum of squares to the mean unexplained sum of squares. These mean explained and unexplained sums of squares are found by dividing each sum of squares by appropriate degrees of freedom, which are k and $N - k - 1$, respectively. N, as always, is the number of cases and k is the number of independent variables. Degrees of freedom in our example, then, are $k = 2$ and $N - k - 1 = 46$. For the R^2 for the regression of fertility rate on percent urban and radio availability, we have:

$$F = \frac{\text{Mean Explained Sum of Squares}}{\text{Mean Unexplained Sum of Squares}}$$

$$= \frac{28.263}{2.438}$$

$$= 11.593$$

We find the probability associated with $F(2, 46) = 11.593$ in Table 3 in the Appendix, just as we did for F values in analysis of variance. Although based on only 49 cases, the multiple correlation for the regression of fertility rate on percent urban and radios per 100 persons is statistically significant at the .001 level. Generalizability to the population is not at issue here since we have population data. But the significance test gives us assurance that the relationship of fertility to the two independent variables is not due to chance processes.

Alternatively, we can find F directly from R^2 with this formula:

$$F = \left(\frac{R^2}{1 - R^2}\right)\left(\frac{N - k - 1}{k}\right)$$

$$= \left(\frac{.335}{1 - .335} \right) \left(\frac{49 - 2 - 1}{2} \right)$$

$$= (.504)(23)$$

$$= 11.592$$

As always, remember that statistical significance depends not only on the strength of a relationship, but also on the number of cases. Even trivial Rs will be found statistically significant if the N is large enough. In our example, however, the significance test is based on a fairly small N (only 49 cases) and thus commands our attention.

12.5   Regression with Dichotomous and Dummy Variables

Multiple regression and correlation can include dichotomous independent variables. Consider the dichotomous variable sex, with males coded 0 and females coded 1.[1] We have already seen (in Section 3.5) that the mean of sex (.56 in the General Social Survey) is the proportion female and (in Section 4.1) that the variance of sex is the proportion male times the proportion female (.56 x .44 = .25). We learned too (in Section 10.8) that we can incorporate dichotomous variables, even nominal level ones like sex, into zero-order regression equations and base correlation coefficients on them.

As an independent variable, sex shows the effect of "femaleness" on a dependent variable in a zero-order or multiple regression. Using General Social Survey data, here is the equation for the regression of occupational prestige (an interval/ratio measure of occupational status) on sex (X_1) and education (X_2):

$$Y' = 8.44 - .22X_1 + 2.56X_2$$

where Y' = predicted prestige score

X_1 = score on sex

X_2 = score on education

We interpret the regression and correlation coefficients for sex just as we do for any variable. The unstandardized regression coefficient of –.22 indicates that females have lower occupational prestige than males, controlling for education. The multiple correlation

---

[1] As noted in Section 10.8, males are actually coded 1 and females are coded 2 in the General Social Survey. However, the logic of dummy variables is the same. Only the actual sizes of coefficients are affected by the codes.

squared is .29, mostly due to the substantial effect of education on occupational prestige. (Incidentally, were the coding of sex reversed, with females coded 0 and males 1, the signs of the zero-order regression and correlation coefficients would change but their magnitudes would remain the same. Coefficients would then show the effect of "maleness" on occupational prestige.)

Inclusion of dichotomous independent variables, even nominal ones, in regression and correlation analyses is easy enough. The "mechanics" are exactly the same as for other variables. But nominal variables with more than two values are a little more complicated to handle. They must be transformed into dummy variables before inclusion in a regression model. A *dummy variable* has only the values of 0 and 1, with 0 indicating the absence of an attribute and 1 indicating its presence.

Dummy variables come in sets of two or more. I will show you how this works using religious preference as an example. Here are the codes for the variable's five values:

1   Protestant
2   Catholic
3   Jewish
4   None
5   Other

We can create four dummy variables that retain information on respondents' religion:

Dummy Variable	Codes
Rel-Cath	1 if Catholic 0 otherwise
Rel-Jewish	1 if Jewish 0 otherwise
Rel-None	1 if None 0 otherwise
Rel-Other	1 if Other 0 otherwise

Catholics are distinguished by code 1 on variable Rel-Cath and 0 on the other dummy variables. Jewish respondents are uniquely identified by 1 on Rel-Jewish and 0 on the other dummy variables. Likewise, Nones and Others are coded 1 on Rel-None and Rel-Other, respectively, and zero on the other variables. Note that we need only *four* dummy variables to capture all of the information about *five* religious preferences since Protestant respondents (the fifth value) are uniquely distinguished by 0 on all four dummy variables. (There are some technical reasons that recommend treating the modal category—Protestant for this variable—as the "reference value" coded 0 on all dummy variables.) A little thought will convince you that we always use one fewer dummy variable than there are values of the original variable. All of this is easy. (Even a dummy can dummy.)

As for other dichotomous variables coded 0 and 1, a dummy variable's mean is the proportion of cases with the characteristics identified by code 1. For the religious affiliation variable dummied above, the mean of the Rel-Cath is the proportion of Catholics, the mean of Rel-Jewish is the proportion of Jews, and so on. We can incorporate these four dummy variables in a regression and correlation analysis. That is, we can regress a dependent variable Y on the variables Rel-Cath, Rel-Jewish, Rel-Other, and Rel-None. Coefficients for each dummy variable tell us the effect of that attribute (e.g., being Catholic) on the dependent variable. These effects are described after controlling for any other variables in the analysis.

Moreover, if a set of dummy variables are used without additional independent variables, the intercept (or constant) and the unstandardized regression coefficients describe dependent variable means for categories distinguished by dummy variables. The intercept is the mean of the category coded 0 on all dummy variables. The intercept plus a dummy variable's unstandardized regression coefficient is the dependent variable mean for the category distinguished by that dummy variable. Table 12.1 presents coefficients for the regression of education on the religion dummy variables:

Table 12.1. Regression of Years of Education on Religion Dummy Variable

Variable	b	Beta	Mean
Rel-Cath	.410	.059	13.510
Rel-Jewish	2.268	.117	15.368
Rel-Other	1.361	.100	14.461
Rel-None	.419	.046	13.519
Constant	13.100		13.100

$R^2 = .02$; $F (3,2887) = 17.309$; $p < .001$

Means are not usually presented as such in a table of regression results like this, but I have listed them on the far right so that you can see that each religious group's education mean is equal to the intercept plus the unstandardized regression coefficient for that group's dummy variable. The last mean listed—13.100—is the average for Protestants, and it equals the Y-intercept or constant. Substantively, by the way, this regression shows the relatively greater effects of Jewish or Other religious identity on educational achievement. Overall, though, the relationship of religion to education is relatively small even though statistically significant. R^2 is .02, indicating that this set of dummy variables for religion explains only about 2 percent of the variation in years of education.

Once again, we see the interweaving of statistical ideas and procedures as we have so often in our study of statistics. Sums of squares reappear in this chapter as regression blends with analysis of variance and as regression coefficients and intercepts describe means. The grace and elegance as well as the rigor and strength of statistics lie in just these sorts of interconnections.

12.6   Summing Up Chapter 12

Here is what we have learned in this chapter:

- Multiple regression extends the bivariate regression model to include additional independent variables.

- With two or more independent variables, the multiple regression model assumes interval/ratio levels of measurement, linear relationships, additive effects (i.e., no interaction), and uncorrelated independent variables.

- A multiple correlation coefficient describes the strength of the relationship between a dependent variable and a set of two or more independent variables.

- The square of the multiple correlation coefficient is the proportion of variation in the dependent variable explained by the set of independent variables considered simultaneously.

- A beta coefficient describes the standard deviation change in the dependent variable associated with an increase of one standard deviation in the independent variable, controlling for the effects of other independent variables.

- Tests of statistical significance for multiple correlation coefficients assume random sampling, normal distributions of the dependent variable within values of the independent variable(s), and equal dependent variable variances within values of the independent variable(s).

- Nominal variables can be incorporated into regression and correlation analyses by first converting them into dummy variables.

- The intercept and unstandardized regression coefficients for dummy variables describe dependent variable means for categories distinguished by the dummy variables.

Key Concepts and Procedures

Ideas and Terms

multiple regression model	multicollinearity
multiple regression equation	multiple correlation coefficient
unstandardized regression coefficients	beta coefficient
statistical interaction	dummy variable

Symbols

Y'

b_1

b_2

$R^2_{Y \cdot 12 \ldots}$

$\beta_{Y1 \cdot 2 \ldots}$

Formulas

$$Y' = a + b_1 X_1 + b_2 X_2$$

$$F = \left(\frac{R^2}{1 - R^2} \right) \left(\frac{N - k - 1}{k} \right)$$

$$R^2 = \frac{\text{Regression Sum of Square}}{\text{Total Sum of Squares}}$$

$$\beta_{Y1 \cdot 2} = \frac{r_{Y1} - r_{Y2} r_{12}}{1 - r_{12}^2}$$

$$F = \frac{\text{Mean Explained Sum of Squares}}{\text{Mean Unexplained Sum of Squares}}$$

$$\beta_{Y2 \cdot 1} = \frac{r_{Y2} - r_{Y1} r_{21}}{1 - r_{21}^2}$$

ANALYSIS WRITE-UP 8: MULTIPLE REGRESSION AND CORRELATION

✦   ✦   ✦   ✦   ✦

As with other multivariate analyses, write-ups of multiple regression and correlation analyses properly begin with univariate descriptions of variables and perhaps descriptions of bivariate relationships among them. Here I have collapsed the univariate write-up and greatly abridged the bivariate write-up in order to focus on reporting multivariate results.

Child abuse is a complex problem with many sources. Economic conditions may affect child abuse. Perhaps education too. Maybe family structure and anomie. With data for the 50 American states, I assessed the possible effects of some of these social variables by regressing child abuse rate on school dropout rate, unemployment rate, church membership percentage, and percent of households male-headed with children but no spouse. Child abuse rate is the number of children involved in a substantial incident of abuse per 1000 children. Listwise deletion of missing data, with N = 48, is used throughout this analysis.

Table 1 reports intercorrelations among variables. Child abuse has modest and statistically insignificant correlations with independent variables ranging from .17 to .30. Relationships among independent variables are generally small and not statistically significant except for moderately strong correlations between dropout and unemployment rates (.56) and between church membership and percent of male-headed households (−.59).

TABLE 1 ABOUT HERE

The multiple regression of child abuse rate onto the five independent variables finds the largest beta for percent of households with children headed only by males (beta = .247). The beta associated with school dropout rate also has a noticeable effect (beta = .117). Neither beta is statistically significant at the .05 level, however, and betas for unemployment and church membership (.086 and −.040, respectively) indicate even smaller effects of those variables. Overall, the four variables taken together explain about 12 percent of the variation in child abuse rates. The F (4, 43) = 1.459 for this multiple regression is not statistically significant.

TABLE 2 ABOUT HERE

Include Tables 1 and 2 at the end of the paper or report:

Table 1. Correlations (r) Among Child Abuse and Independent Variables

Variable	Child Abuse	Dropout Rate	Unempl'nt Rate	Church Mmbrshp	% Hshlds Male-Head
Child Abuse Rate	1.00	.20	.17	−.18	.30*
School Dropout Rate		1.00	.56**	.01	.16
Unemployment Rate			1.00	.00	.08
Church Membership				1.00	−.59**
% Hshlds Male-Headed					1.00

* p < .05; ** p < .01

Table 2. Regression of Child Abuse Rate on Social and Economic Variables

Variable	b	Beta
School Dropout Rate	.115	.117
Unemployment Rate	.468	.086
Church Membership	−.025	−.040
% Hshlds Male-Headed	5.873	.247
Constant	3.518	

R^2 = .12; F (4,43) = 1.459; n.s.

Appendix
Statistical Tables

TABLE 1
The Chi-Square Distribution

TABLE 2
The *t* Distribution

TABLE 3
The *F* Distribution

TABLE 1: THE CHI-SQUARE DISTRIBUTION

Probability

df	.05	.02	.01	.001
1	3.841	5.412	6.635	10.827
2	5.991	7.824	9.210	13.815
3	7.815	9.837	11.345	16.266
4	9.488	11.668	13.277	18.467
5	11.070	13.388	15.086	20.515
6	12.592	15.033	16.812	22.457
7	14.067	16.622	18.475	24.322
8	15.507	18.168	20.090	26.125
9	16.919	19.679	21.666	27.877
10	18.307	21.161	23.209	29.588
11	19.675	22.618	24.725	31.264
12	21.026	24.054	26.217	32.909
13	22.362	25.472	27.688	34.528
14	23.685	26.873	29.141	36.123
15	24.996	28.259	30.578	37.697
16	26.296	29.633	32.000	39.252
17	27.587	30.995	33.409	40.790
18	28.869	32.346	34.805	42.312
19	30.144	33.687	36.191	43.820
20	31.410	35.020	37.566	45.315
21	32.671	36.343	38.932	46.797
22	33.924	37.659	40.289	48.268
23	35.172	38.968	41.638	49.728
24	36.415	40.270	42.980	51.179
25	37.652	41.566	44.314	52.620
26	38.885	42.856	45.642	54.052
27	40.113	44.140	46.963	55.476
28	41.337	45.419	48.278	56.893
29	42.557	46.693	49.588	58.302
30	43.773	47.962	50.892	59.703

Source: Adapted from Table IV of Fisher and Yates, *Statistical Tables for Biological, Agricultural, and Medical Research*, published by Longman Group UK Ltd., 1974. Reprinted with permission of the authors and publishers.

TABLE 2: THE t DISTRIBUTION

Level of Significance for One-Tailed Test

	.10	.05	.025	.01	.005	.0005
	Level of Significance for Two-Tailed Test					
	.20	.10	.05	.02	.01	.001
1	3.078	6.314	12.706	31.821	63.657	636.619
2	1.886	2.920	4.303	6.965	9.925	31.598
3	1.638	2.353	3.182	4.451	5.841	12.941
4	1.533	2.132	2.776	3.747	4.604	8.610
5	1.476	2.015	2.571	3.365	4.032	6.859
6	1.440	1.943	2.447	3.143	3.707	5.959
7	1.415	1.895	2.365	2.998	3.499	5.405
8	1.397	1.860	2.306	2.896	3.355	5.041
9	1.383	1.833	2.262	2.821	3.250	4.781
10	1.372	1.812	2.228	2.764	3.169	4.587
11	1.363	1.796	2.201	2.718	3.106	4.437
12	1.356	1.782	2.179	2.681	3.055	4.318
13	1.350	1.771	2.160	2.650	3.012	4.221
14	1.345	1.761	2.145	2.624	2.977	4.140
15	1.341	1.753	2.131	2.602	2.947	4.073
16	1.337	1.746	2.120	2.583	2.921	4.015
17	1.333	1.740	2.110	2.567	2.898	3.965
18	1.330	1.734	2.101	2.552	2.878	3.922
19	1.328	1.729	2.093	2.539	2.861	3.883
20	1.325	1.725	2.086	2.528	2.845	3.850
21	1.323	1.721	2.080	2.518	2.831	3.819
22	1.321	1.717	2.074	2.508	2.819	3.792
23	1.319	1.714	2.069	2.500	2.807	3.767
24	1.318	1.711	2.064	2.492	2.797	3.745
25	1.316	1.708	2.060	2.485	2.787	3.725
26	1.315	1.706	2.556	2.479	2.779	3.707
27	1.314	1.703	2.052	2.473	2.771	3.690
28	1.313	1.701	2.048	2.467	2.763	3.674
29	1.311	1.699	2.045	2.462	2.756	3.659
30	1.310	1.697	2.042	2.457	2.750	3.646
40	1.303	1.684	2.021	2.423	2.704	3.551
60	1.296	1.671	2.000	2.390	2.660	3.460
120	1.289	1.658	1.980	2.358	2.617	3.373
∞	1.282	1.645	1.960	2.326	2.576	3.291

Source: Adapted from Table III of Fisher and Yates, *Statistical Tables for Biological, Agricultural, and Medical Research*, published by Longman Group UK Ltd., 1974. Reprinted with permission of the authors and publishers.

TABLE 3A: THE *F* DISTRIBUTION

$$p = .05$$

N_2 \ N_1	1	2	3	4	5	6	8	12	24	∞
1	161.4	199.5	215.7	224.6	230.2	234.0	238.9	243.9	249.0	254.3
2	18.51	19.00	19.16	19.25	19.30	19.33	19.37	19.41	19.45	19.50
3	10.13	9.55	9.28	9.12	9.01	8.94	8.84	8.74	8.64	8.53
4	7.71	6.94	6.59	6.39	6.26	6.16	6.04	5.91	5.77	5.63
5	6.61	5.79	5.41	5.19	5.05	4.95	4.82	4.68	4.53	4.36
6	5.99	5.14	4.76	4.53	4.39	4.28	4.15	4.00	3.84	3.67
7	5.59	4.74	4.35	4.12	3.97	3.87	3.73	3.57	3.41	3.23
8	5.32	4.46	4.07	3.84	3.69	3.58	3.44	3.28	3.12	2.93
9	5.12	4.26	3.86	3.63	3.48	3.37	3.23	3.07	2.90	2.71
10	4.96	4.10	3.71	3.48	3.33	3.22	3.07	2.91	2.74	2.54
11	4.84	3.98	3.59	3.36	3.20	3.09	2.95	2.79	2.61	2.40
12	4.75	3.88	3.49	3.26	3.11	3.00	2.85	2.69	2.50	2.30
13	4.67	3.80	3.41	3.18	3.02	2.92	2.77	2.60	2.42	2.21
14	4.60	3.74	3.34	3.11	2.96	2.85	2.70	2.53	2.35	2.13
15	4.54	3.68	3.29	3.06	2.90	2.79	2.64	2.48	2.29	2.07
16	4.49	3.63	3.24	3.01	2.85	2.74	2.59	2.42	2.24	2.01
17	4.45	3.59	3.20	2.96	2.81	2.70	2.55	2.38	2.19	1.96
18	4.41	3.55	3.16	2.93	2.77	2.66	2.51	2.34	2.15	1.92
19	4.38	3.52	3.13	2.90	2.74	2.63	2.48	2.31	2.11	1.88
20	4.35	3.49	3.10	2.87	2.71	2.60	2.45	2.28	2.08	1.84
21	4.32	3.47	3.07	2.84	2.68	2.57	2.42	2.25	2.05	1.81
22	4.30	3.44	3.05	2.82	2.66	2.55	2.40	2.23	2.03	1.78
23	4.28	3.42	3.03	2.80	2.64	2.53	2.38	2.20	2.00	1.76
24	4.26	3.40	3.01	2.78	2.62	2.51	2.36	2.18	1.98	1.73
25	4.24	3.38	2.99	2.76	2.60	2.49	2.34	2.16	1.96	1.71
26	4.22	3.37	2.98	2.74	2.59	2.47	2.32	2.15	1.95	1.69
27	4.21	3.35	2.96	2.73	2.57	2.46	2.30	2.13	1.93	1.67
28	4.20	3.34	2.95	2.71	2.56	2.44	2.29	2.12	1.91	1.65
29	4.18	3.33	2.93	2.70	2.54	2.43	2.28	2.10	1.90	1.64
30	4.17	3.32	2.92	2.69	2.53	2.42	2.27	2.09	1.89	1.62
40	4.08	3.23	2.84	2.61	2.45	2.34	2.18	2.00	1.79	1.51
60	4.00	3.15	2.76	2.52	2.37	2.25	2.10	1.92	1.70	1.39
120	3.92	3.07	2.68	2.45	2.29	2.17	2.02	1.83	1.61	1.25
∞	3.84	2.99	2.60	2.37	2.21	2.10	1.94	1.75	1.52	1.00

Values of N_1 and N_2 are the degrees of freedom associated with the larger and smaller estimates of variance, respectively.

Adapted from Table IV of Fisher and Yates, *Statistical Tables for Biological, Agricultural, and Medical Research*, published by Longman Group UK Ltd., 1974. Reprinted with permission of the authors and publishers.

TABLE 3B: THE *F* DISTRIBUTION (Continued)

$p = .01$

N_2 \ N_1	1	2	3	4	5	6	8	12	24	∞
1	4052	4999	5403	5625	5764	5859	5982	6106	6234	6336
2	98.50	99.00	99.17	99.25	99.30	99.33	99.37	99.42	99.46	99.50
3	34.12	30.82	29.46	28.71	28.24	27.91	27.49	27.05	26.60	26.12
4	21.20	18.00	16.69	15.98	15.52	15.21	14.80	14.37	13.93	13.46
5	16.26	13.27	12.06	11.39	10.97	10.67	10.29	9.89	9.47	9.02
6	13.74	10.92	9.78	9.15	8.75	8.47	8.10	7.72	7.31	6.88
7	12.25	9.55	8.45	7.85	7.46	7.19	6.84	6.47	6.07	5.65
8	11.26	8.65	7.59	7.01	6.63	6.37	6.03	5.67	5.28	4.86
9	10.56	8.02	6.99	6.42	6.06	5.80	5.47	5.11	4.73	4.31
10	10.04	7.56	6.55	5.99	5.64	5.39	5.06	4.71	4.33	3.91
11	9.65	7.20	6.22	5.67	5.32	5.07	4.74	4.40	4.02	3.60
12	9.33	6.93	5.95	5.41	5.06	4.82	4.50	4.16	3.78	3.36
13	9.07	6.70	5.74	5.20	4.86	4.62	4.30	3.96	3.59	3.16
14	8.86	6.51	5.56	5.03	4.69	4.46	4.14	3.80	3.43	3.00
15	8.68	6.36	5.42	4.89	4.56	4.32	4.00	3.67	3.29	2.87
16	8.53	6.23	5.29	4.77	4.44	4.20	3.89	3.55	3.18	2.75
17	8.40	6.11	5.18	4.67	4.34	4.10	3.79	3.45	3.08	2.65
18	8.28	6.01	5.09	4.58	4.25	4.01	3.71	3.37	3.00	2.57
19	8.18	5.93	5.01	4.50	4.17	3.94	3.63	3.30	2.92	2.49
20	8.10	5.85	4.94	4.43	4.10	3.87	3.56	3.23	2.86	2.42
21	8.02	5.78	4.87	4.37	4.04	3.81	3.51	3.17	2.80	2.36
22	7.94	5.72	4.82	4.31	3.99	3.76	3.45	3.12	2.75	2.31
23	7.88	5.66	4.76	4.26	3.94	3.71	3.41	3.07	2.70	2.26
24	7.82	5.61	4.72	4.22	3.90	3.67	3.36	3.03	2.66	2.21
25	7.77	5.57	4.68	4.18	3.86	3.63	3.32	2.99	2.62	2.17
26	7.72	5.53	4.64	4.14	3.82	3.59	3.29	2.96	2.58	2.13
27	7.68	5.49	4.60	4.11	3.78	3.56	3.26	2.93	2.55	2.10
28	7.64	5.45	4.57	4.07	3.75	3.53	3.23	2.90	2.52	2.06
29	7.60	5.42	4.54	4.04	3.73	3.50	3.20	2.87	2.49	2.03
30	7.56	5.39	4.51	4.02	3.70	3.47	3.17	2.84	2.47	2.01
40	7.31	5.18	4.31	3.83	3.51	3.29	2.99	2.66	2.29	1.80
60	7.08	4.98	4.13	3.65	3.34	3.12	2.82	2.50	2.12	1.60
120	6.85	4.79	3.95	3.48	3.17	2.96	2.66	2.34	1.95	1.38
∞	6.64	4.60	3.78	3.32	3.02	2.80	2.51	2.18	1.79	1.00

Values of N_1 and N_2 are the degrees of freedom associated with the larger and smaller estimates of variance, respectively.

TABLE 3C: THE *F* DISTRIBUTION (Continued)

$p = .001$

N_2 \ N_1	1	2	3	4	5	6	8	12	24	∞
1	405284	500000	540379	562500	576405	585937	598144	610667	623497	636619
2	998.5	999.0	999.2	999.2	999.3	999.3	999.4	999.4	999.5	999.5
3	167.0	148.5	141.1	137.1	134.6	132.8	130.6	128.3	125.9	123.5
4	74.14	61.25	56.18	53.44	51.71	50.53	49.00	47.41	45.77	44.05
5	47.18	37.12	33.20	31.09	29.75	28.84	27.64	26.42	25.14	23.78
6	35.51	27.00	23.70	21.92	20.81	20.03	19.03	17.99	16.89	15.75
7	29.25	21.69	18.77	17.19	16.21	15.52	14.63	13.71	12.73	11.69
8	25.42	18.49	15.83	14.39	13.49	12.86	12.04	11.19	10.30	9.34
9	22.86	16.39	13.90	12.56	11.71	11.13	10.37	9.57	8.72	7.81
10	21.04	14.91	12.55	11.28	10.48	9.92	9.20	8.45	7.64	6.76
11	19.69	13.81	11.56	10.35	9.58	9.05	8.35	7.63	6.85	6.00
12	18.64	12.97	10.80	9.63	8.89	8.38	7.71	7.00	6.25	5.42
13	17.81	12.31	10.21	9.07	8.35	7.86	7.21	6.52	5.78	4.97
14	17.14	11.78	9.73	8.62	7.92	7.43	6.80	6.13	5.41	4.60
15	16.59	11.34	9.34	8.25	7.57	7.09	6.47	5.81	5.10	4.31
16	16.12	10.97	9.00	7.94	7.27	6.81	6.19	5.55	4.85	4.06
17	15.72	10.66	8.73	7.68	7.02	6.56	5.96	5.32	4.63	3.85
18	15.38	10.39	8.49	7.46	6.81	6.35	5.76	5.13	4.45	3.67
19	15.08	10.16	8.28	7.26	6.62	6.18	5.59	4.97	4.29	3.52
20	14.82	9.95	8.10	7.10	6.46	6.02	5.44	4.82	4.15	3.38
21	14.59	9.77	7.94	6.95	6.32	5.88	5.31	4.70	4.03	3.26
22	14.38	9.61	7.80	6.81	6.19	5.76	5.19	4.58	3.92	3.15
23	14.19	9.47	7.67	6.69	6.08	5.65	5.09	4.48	3.82	3.05
24	14.03	9.34	7.55	6.59	5.98	5.55	4.99	4.39	3.74	2.97
25	13.88	9.22	7.45	6.49	5.88	5.46	4.91	4.31	3.66	2.89
26	13.74	9.12	7.36	6.41	5.80	5.38	4.83	4.24	3.59	2.82
27	13.61	9.02	7.27	6.33	5.73	5.31	4.76	4.17	3.52	2.75
28	13.50	8.93	7.19	6.25	5.66	5.24	4.69	4.11	3.46	2.70
29	13.39	8.85	7.12	6.19	5.59	5.18	4.64	4.05	3.41	2.64
30	13.29	8.77	7.05	6.12	5.53	5.12	4.58	4.00	3.36	2.59
40	12.61	8.25	6.60	5.70	5.13	4.73	4.21	3.64	3.01	2.23
60	11.97	7.76	6.17	5.31	4.76	4.37	3.87	3.31	2.69	1.90
120	11.38	7.32	5.79	4.95	4.42	4.04	3.55	3.02	2.40	1.54
∞	10.83	6.91	5.42	4.62	4.10	3.74	3.27	2.74	2.13	1.00

Values of N_1 and N_2 are the degrees of freedom associated with the larger and smaller estimates of variance, respectively.

Glossary

Aggregate Data or Variables

Data or variables based on groupings of individual units.

Alpha Error

Rejection of a null hypothesis that is true. Also called a Type I error.

Alpha Level (α)

The probability of rejecting a null hypothesis that is true. Also called significance level.

Analysis of Variance (ANOVA)

Bivariate technique for analyzing differences among means.

ANOVA Table

Table presenting the results of an analysis of variance.

Antecedent Variable

Variable that occurs causally prior to the independent variable (and thus also prior to the dependent variable).

Area Map

Map displaying spatial distribution of an ecological variable.

Association

Relationship between two variables.

Asymmetric Measure of Association

Measure of the strength of a relationship that distinguishes between the independent and dependent variables.

Average (Measure of Central Tendency)

Typical or representative value for a set of scores. (See Mean, Median, Mode.)

Bar Graph (Histogram)

Graph of the distribution of a variable that shows relative frequencies or percentages by the heights of bars. (See Histogram.)

Beta Coefficient (Beta-Weight or Standardized Regression Coefficient)

The standard deviation change in the dependent variable associated with a one standard deviation increase in an independent variable, controlling for other independent variables. Same as standardized regression coefficient.

Beta Error

Failure to reject a null hypothesis that is false. Also called Type II error.

Between-Groups Sum of Squares

A measure of variation that is the weighted sum of the squared deviations of group means from the total mean.

Bimodal Distribution

Distribution in which two values occur much more often than other values (i.e., a distribution with two modes).

Bivariate Analysis

Analysis of the relationship between two variables.

Box-and-Whisker Diagram (Boxplot)

Graphic representation of the distributions of dependent variable scores within categories of the independent variable.

C (Pearson's Contingency Coefficient)

Chi square–based, symmetric measure of association for nominal variables.

Causal Relationship

Relationship between two variables that meets the causal conditions of temporal ordering, association, and nonspuriousness.

Cell Frequency

Number in a table indicating count of scores with a given value or joint values.

Central Limit Theorem

As the sample size N increases, the sampling distribution of the mean more and more closely resembles a normal distribution, with a mean equal to the population mean and a standard deviation of $\frac{\sigma}{\sqrt{N}}$.

Chi Square (χ^2)

Statistic that compares the actual frequencies in a bivariate table with the frequencies expected if there is no relationship between the variables. Used for tests of statistical significance and for some measures of association in cross-tabulation.

Codebook
Listing of information about variables in a data set.

Codes
Symbols (usually numerals) that indicate cases' scores on variables.

Collectively Exhaustive
Condition in which values of a variable include all possible cases.

Confidence Interval
The range of values of a sample statistic within which a population parameter lies with a specified probability.

Constant (Y-Intercept)
In regression, value of the dependent variable at which a regression line or plane crosses the Y axis.

Control Variable (Test Factor)
Variable held constant while examining a bivariate relationship.

Correlation Coefficient (r)
Measure of association between two interval/ratio variables indicating the strength and direction of their relationship; summary measure of the extent to which cases are clustered about the regression line.

Correlation Matrix
Square array of correlation coefficients.

Cross-product
The product of deviations of scores from their means.

Cross-tabulation
Analysis of the associations among variables by comparing percentage distributions.

Cumulative Frequency (F)
Number of scores that have a given value or less.

Cumulative Percentage
Percentage of scores that have a given value or less.

Curvilinear Relationship
Relationship between two ordinal or interval/ratio variables that changes direction.

Data
Records of observations.

Data File

Data set stored in a form that can be used by a computer.

Data Set

Data organized in a systematic way.

Degrees of Freedom (df)

Number of scores or cell frequencies free to vary when computing a statistic; used with statistics such as chi square and the F ratio for significance tests.

Dependent Variable

Variable that is caused by an independent variable.

Descriptive Statistics

Methods for quantitatively summarizing information.

Dichotomous Variable

Variable with only two values.

Dummy Variable

Variable with values 0 and 1 used to introduce nominal variables into regression and correlation analyses.

D_{YX} and D_{XY}

See Somers' D_{YX} and D_{XY}.

E^2 (Eta Squared)

Nonlinear measure of association that describes the proportion of variation in an interval/ratio dependent variable explained by a categorical independent variable.

Ecological Data or Variables

Aggregate data or variables based on spatial or geographic units such as city districts, states, or countries.

Elaboration

Process of introducing controls for antecedent or intervening variables.

Explained Variation

Variation in dependent variable scores explained by independent variable scores.

Explanation

Demonstration that a bivariate association is spurious (i.e., accounted for by an antecedent variable).

f (Frequency)

Number of cases with a particular value of a variable or joint values of two or more variables.

f_e (Expected Frequency)
Number of cases in a cell that would be expected if the two variables were unrelated.

f_o (Observed Frequency)
Actual number of cases in a cell of a table.

F (F Ratio)
Statistic formed by the ratio of between-groups variance to within-groups variance. Used in tests of significance.

First-Order Partial Correlation
Partial correlation with one control variable.

First-Order Partial Table
Subtable of a partial table with only one control variable.

Frequency Distribution
Array of numbers of cases with each value of a variable or joint values of two or more variables.

Gamma (G or γ)
Symmetric PRE measure of association for ordinal variables.

General Social Survey
Annual nationwide survey carried out since 1972 by the National Opinion Research Center.

Genuine Relationship
A causal association that does not appear to be explained by an antecedent variable.

Histogram
Type of bar graph with contiguous bars for continuous variables.

Homoscedasticity
Condition in which distributions of dependent variable within independent variable categories have equal variances.

Inferential Statistics
Methods for generalizing from a sample to the larger population from which the sample was drawn.

Interpretation
Demonstration that an intervening variable links an independent and dependent variable.

Interval/Ratio Variable
Variable measured using a standard or fixed unit of measurement.

Intervening Variable

Variable that causally links an independent and a dependent variable.

Joint Frequency

Number of cases with given values on two or more variables.

Kurtosis

Peakedness of a distribution.

Lambda (λ)

Asymmetric PRE measure of association for nominal variables.

Level of Measurement

Classification of ways in which variables are measured (i.e., nominal variable, ordinal variable, interval/ratio variable).

Level of Significance

Probability that a relationship found in sample data occurs by chance if there is no relationship in the population.

Linear Equation

An equation for a best-fitting straight line or plane that describes the relationship between/among variables.

Linear Regression

Describing a relationship between/among variables with a best-fitting straight line or plane.

Listwise Deletion

Exclusion of cases missing information on any of the variables used in an analysis, even if not missing information for the particular statistic being calculated.

Major Diagonal

Diagonal of a table running from the upper left to lower right ($\backslash$).

Marginal Distributions

Row totals and column totals of a frequency table.

Mean ($\overline{X}$ or μ)

Arithmetical average of all scores; the sum of cases divided by the number of cases.

Mean Square

Measure of variation calculated by dividing a sum of squares by corresponding degrees of freedom.

Measure of Association

Statistic summarizing the strength (and sometimes the direction) of a relationship.

Measure of Variation
Summary of how spread out scores are. (See Standard Deviation; Variance.)

Measurement
Process of finding cases' scores on a variable.

Median (Md)
Value that divides an ordered set of scores in half.

MicroCase
Handy computer program for statistics.

Minor Diagonal
Diagonal of a table running from the lower left to upper right (/).

Missing Data
Values of a variable that provide little or no information for analysis. Usually includes values such as Don't Know, No Opinion, No Answer, Refused, Not Applicable, and Not Ascertained.

Mode (Mo)
Most frequently occurring score on a variable.

Multicollinearity
Situation in which correlations between independent variables are large.

Multiple Correlation Coefficient (R)
Measure of the strength of the combined effect of two or more independent variables on a dependent variable.

Multiple Regression Equation
Mathematical expression that describes the relationship of a dependent variable to two or more independent variables.

Multivariate Analysis
Simultaneous analysis of data for three or more variables.

Mutually Exclusive
Condition in which each case has only one value on a variable (i.e., values do not overlap).

N
Number of cases in an analysis.

Negative Relationship
Relationship in which higher scores on one variable are associated with lower scores on the other variable.

Nominal Variable

Variable with values that are unordered categories.

Normal Distribution

Mathematically defined, bell-shaped distribution in which known proportions of scores lie between the mean and specified standard deviations.

Null Hypothesis (H_0)

Assertion or assumption of no relationship between variables in the population.

One-Way Analysis of Variance

Analysis of variance involving a single independent variable.

Ordinal Variable

Variable with values that can be rank-ordered but that are not measured with a fixed unit of measurement.

Outlier

Variable score that is unusually low or high.

ϕ (Phi)

Chi square–based nominal measure of association for 2 by c tables or r by 2 tables; equivalent to V in 2 by 2 tables.

Pairwise Deletion

Exclusion only of cases missing information on the particular variables used in the calculation of a statistic.

Parameter

Characteristic of a population. (See Statistic.)

Partial Gamma (G_P)

Measure of association for multivariate tabular analysis of ordinal variables.

Partial Table

Table for each category of control variable(s).

Percentage

Standardized frequency assuming a total of 100 cases.

Pie Chart

Graphic representation of the distribution of a variable that shows segments of a circle proportional in area to the frequencies or percentages.

Playing

Along with thinking, the most important activity in statistical analysis.

Population
Set of cases from which a sample is drawn and to which a researcher wants to generalize. (See Sample.)

Positive Relationship
Relationship in which higher scores on one variable are associated with higher scores on the other variable.

Proportionate Reduction in Error (PRE) Measures
Measures of association that report the proportion by which errors in predicting dependent variable scores are reduced by knowledge of independent variable scores.

r (Pearson Product-Moment Correlation Coefficient)
Measure of association for the relationship between two interval/ratio variables.

r^2 (Coefficient of Determination)
Proportion of variation in a dependent variable explained by an independent variable.

R (Multiple Correlation Coefficient)
Measure of association for the relationship of a dependent variable to two or more independent variables.

R^2 (Coefficient of Multiple Determination)
Proportion of variation in a dependent variable explained by two or more independent variables.

Raw Data
Initial observations or scores in a data set.

Regression Equation
Mathematical expression that describes the regression line.

Regression Line (Least-Squares Line)
Summary line on a scatterplot that minimizes the sum of squares of residuals.

Replication
Demonstration that a bivariate relationship is largely unaffected by control(s) for antecedent variable(s).

Research Hypothesis
Expectation about the relationship between variables.

Residual
Error left in predicting a dependent variable score from an independent variable score.

Sample

Set of cases taken from a larger population of cases. (See Population.)

Sampling Distribution

Distribution of some sample statistic for all possible samples of a certain size drawn from a particular population.

Scatterplot (Scattergram)

Graphic representation of relationships between interval/ratio variables.

Score

Case's value on a variable.

Significant Digits

Digits that are reliable (e.g., in percentages or measures of association).

Somers' D_{YX} and D_{XY}

Asymmetric measures of association for ordinal variables that take dependent variable ties into account.

Specification

Demonstration that a bivariate relationship varies depending on the value of a control variable.

Spot Map

Map displaying the spatial distribution of an ecological variable using dots whose sizes indicate spatial unit scores on the variable.

Spurious Relationship

Statistical association that is not a causal relationship but instead is due to some antecedent variable(s).

Stacked Bar Graph

Bivariate bar graph in which segments of independent variable bars correspond in length to frequencies or percentages of designated dependent variables.

Standard Deviation (s or σ)

Measure of variation in scores; square root of the variance. (See Variance.)

Standard Error

The standard deviation of a sampling distribution.

Standardized Regression Coefficient (Beta Coefficient or Beta-Weight)

The standard deviation change in the dependent variable associated with a one standard deviation increase in an independent variable, controlling for other independent variables.

Standardized Variable

Variable whose scores have all been converted to Z-scores.

Statistic

Characteristic of a sample. (See Parameter.)

Statistical Interaction

Condition in which the combined effect of two independent variables is not equal to the sum of their individual effects.

Statistics

1. Numbers that summarize information. 2. Methods for quantitatively summarizing and generalizing information. 3. Characteristics of a sample.

Subset

Cases selected for an analysis on the basis of their scores on one or more specified variables.

Sum of Squares

Measure of variation that sums the squared deviations from a mean.

Suppressor Variable

Variable that masks a zero-order association between two variables.

Symmetric Measure of Association

Measure of association that does not depend on which variable is the dependent variable and which variable is the independent variable.

t Test

Test of statistical significance of the difference between two means. Equivalent to an analysis of variance with an F test for a dichotomous independent variable.

tau-b (τ_b)

Symmetric measure of association for ordinal variables.

tau-c (τ_c)

Symmetric measure of association for ordinal variables.

Table Elaboration

Process of introducing control variables in cross-tabulation.

Thinking

Along with playing, the most important activity in statistical analysis . . . or almost anything else.

Total Sum of Squares

Measure of variation that sums the squared deviations of all scores from the total mean.

Two-Way Analysis of Variance

Analysis of variance involving two independent variables.

Type I Error

Rejection of a null hypothesis that is true. Also called alpha error.

Type II Error

Failure to reject a null hypothesis that is false. Also called beta error.

Unimodal Distribution

Distribution in which one score occurs considerably more often than other scores (i.e., a distribution with one mode).

Univariate Analysis

Analysis of data concerning only one variable.

V (Cramer's V)

Chi square–based, symmetric measure of association for nominal variables.

Variable

Characteristic or property that differs in value from case to case.

Variance

Average squared deviation of scores from the mean. (See Standard Deviation.)

Within-Groups Sum of Squares

Measure of variation that sums the squared deviations of individual scores from group means.

X-Axis

Horizontal axis, usually used for an independent variable.

Y-Axis

Vertical axis, usually used for a dependent variable.

Y-Intercept (Constant)

In regression, value of the dependent variable at which a regression line or plane crosses the Y-axis.

Z-Score (Standard Score)

Standardized score describing how many standard deviations from the mean a score is located.

Zero-Order

Referring to a bivariate relationship (i.e., no control variable).

Zero-Order Correlation

Bivariate correlation coefficient.

Bibliography

Here are some books on statistics and related topics that my students and I have found especially useful.

General Research Methods

Earl Babbie's *Practice of Social Research*, Eighth Edition (Belmont, CA: Wadsworth, 1997) is a widely used introductory research methods text in sociology. Babbie's text is clear and, while emphasizing survey methods, covers most major research techniques used in the social sciences. I also highly recommend Rodney Stark et. al.'s *Contemporary Social Research Methods*, Second Edition (Bellevue, WA: MicroCase Corporation, 1998) which uses a similar approach to this textbook/workbook package. There are other good methods texts, including Royce Singleton, Jr., et al.'s *Approaches to Social Research*, Second Edition) (New York: Oxford, 1993) and Duane R. Monette et al.'s *Applied Social Research*, Third Edition (Fort Worth, TX: Harcourt Brace, 1994).

General Statistics Textbooks

If you need to brush up on elementary mathematics—handling fractions, formulas, graphs, and the like—turn to Andrew R. Baggaley's *Mathematics for Introductory Statistics* (New York: Wiley, 1969), good enough still to be reprinted after more than 25 years. You might find W. Paul Vogt's *Dictionary of Statistics and Methodology* (Newbury Park: Sage, 1993) a handy reference.

You may well find another statistics book helpful as you use this one. Different approaches help clarify statistical reasoning or procedures. I noted in the Preface that there are many good statistics texts. One of the best is Herman J. Loether and Donald G. McTavish's

Descriptive and Inferential Statistics (Boston: Allyn and Bacon, 1993). Loether and McTavish go into much more depth than most introductory texts, including this one. I am also fond of Hubert M. Blalock's *Social Statistics*, Revised Second Edition (New York: McGraw-Hill, 1979), perhaps because I cut my statistical teeth on its first edition. Lawrence C. Hamilton's excellent *Modern Data Analysis* (Pacific Grove, CA: Brooks/Cole, 1990) emphasizes exploratory data analysis. Also very good are Linton Freeman's *Elementary Applied Statistics* (New York: Wiley, 1968), David S. Moore and George P. McCabe's *Introduction to the Practice of Statistics* (New York: Freeman, 1993), and Jack Levin and James Alan Fox's *Elementary Statistics in Social Research*, Fifth Edition (New York: Longman, 1997). Bernhardt Lieberman's edited *Contemporary Problems in Statistics* (New York: Oxford, 1971) is a useful collection of readings on statistics and statistical issues.

Sage Publications of Newbury Park, California, publishes inexpensive paperback books on statistics and research methods in its Quantitative Applications in the Social Sciences series. The series numbered 117 when I last looked, but it is always growing. Most of the books are quite short[1] and each deals with a specific topic. A few titles suggest the range of statistical works in the Sage series: *Ecological Inference; Analysis of Nominal Data; Tests of Significance; Measures of Association; Applied Regression;* and *Multiple Regression in Practice.* You will find the Sage series helpful for specific topics.

There are numerous "popular" treatments of statistics. Among the better—and more fun—are R. Hooke's *How to Tell the Liars from the Statisticians* (New York: Dekker, 1983), M. Hollander and F. Proschan's *Statistical Exorcist: Dispelling Statistics Anxiety* (New York: Dekker, 1984), and, of course, a couple classics: D. Huff's *How to Lie with Statistics* (New York: Norton, 1954) and Hans Zeisel's *Say It With Figures* (New York: Harper and Row, 1957).

For stimulating and provocative appreciation of the mode and standard deviation, be sure to read Stephen Jay Gould's *Full House* (New York: Harmony Books, 1996). Gould uses the mode and standard deviation to explain why there are no longer .400 hitters in baseball and why there is no progress in evolution despite increasingly complex life forms. What a read!

On probability, see Irving Adler's *Probability and Statistics for Everyman: How to Use and Understand the Laws of Chance* (New York: New American Library, 1963). Adler's work is readable and full of interesting examples.

Rather watch statistics on TV? Then pop the Annenberg/Corporation for Public Broadcasting's *Against All Odds: Inside Statistics* into

---

[1] Sage advertisements report that books in the Quantitative Applications series "average" 88 pages, but do not indicate if this average is a mean or median.

your VCR. This is an engaging and well-done 26-part telecourse. Insight Media of New York City distributes over 30 videotapes on statistics and the visual display of information, including *Against All Odds* tapes. Short (20–25 minutes) tapes on verifying basic statistics are also available in the *Why Use Statistics?* series from Films for the Humanities and Sciences.

Tabular Analysis

Several books in the Sage series are very helpful on cross-tabulation: H. T. Reynolds' *Analysis of Nominal Data*, David Hildebrand et al.'s *Analysis of Ordinal Data*, and Albert M. Liebetrau's *Measures of Association*.

On significance tests, see Lawrence B. Mohr's *Understanding Significance Testing* and Ramon E. Henkel's *Tests of Significance*, both in the Sage series. Statisticians have debated the appropriateness and usefulness of significance tests for years. Denton Morrison and Ramon Henkel have compiled arguments pro and con in their *Significance Test Controversy: A Reader* (Chicago: Aldine-Atherton, 1970).

Morris Rosenberg's *Logic of Survey Analysis* (New York: Basic Books, 1968) is the best single work on the elaboration model, although James A. Davis' *Elementary Survey Analysis* (Englewood Cliffs, NJ: Prentice-Hall, 1970) is also excellent. In their *Delinquency Research: An Appraisal of Analytic Methods* (New York: Free Press, 1967), Travis Hirschi and Hanan Selvin raise important issues about causal analysis and the elaboration model. It is insightful even for those of us not particularly interested in delinquency research.

Graphs and Other Visual Displays

No one has written about communicating information with more intelligence or better aesthetic sense than Edward R. Tufte in his *Visual Display of Quantitative Information* (Cheshire, CT: Graphics Press, 1983). This book is a delight for mind and eye. So too is Tufte's broader treatment in *Envisioning Information* (Cheshire, CT: Graphics Press, 1990). These are splendid books and my personal favorites.

Far more prosaic but still useful is Robert Lefferts' *How to Prepare Charts and Graphs for Effective Reports* (New York: Barnes and Noble, 1981) and, at a more advanced level, C. F. Schmid's *Statistical Graphics: Design Principles and Practices* (Melbourne, FL: Krieger, 1992). William S. Cleveland's *Elements of Graphing Data* (Murray Hill, NJ: AT&T Bell Laboratories, 1994) is also very good.

Analysis of Variance, Regression, and Correlation

You may find Larry D. Schroeder et al.'s *Understanding Regression Analysis* and Michael S. Lewis-Beck's *Applied Regression* helpful introductions to those topics. See William D. Berry and Stanley Feldman's *Multiple Regression in Practice* and Gudman R. Iversen and Helmut Norpoth's *Analysis of Variance*, Second Edition, for more advanced treatments. All of these works are in the Sage Quantitative Applications series. W. C. Guenther's *Analysis of Variance* (Englewood Cliffs, NJ: Prentice-Hall, 1964) and Mordecai Ezekiel and K. A. Fox's *Methods of Correlation and Regression Analysis*, Third Edition (New York: Wiley, 1967) are standard references.

Miscellaneous Works

The American Sociological Association publishes a handy little *Style Guide* whose value lies in its brevity and no-nonsense clarity. Lee Cuba's *Short Guide to Writing About Social Science* (New York: HarperCollins, 1993), the *Sociology Writing Group's Guide to Writing Sociology Papers* (New York: St. Martin's Press, 1994), and Fred Pyrczak and Randall R. Bruce's *Writing Empirical Research Reports* (Los Angeles: Pyrczak, 1996) usefully discuss writing strategies as well as style and format.

Curious about the history of statistics? Stephen M. Stigler's *History of Statistics: The Measurement of Uncertainty Before 1900* (Cambridge, MA: Harvard, 1986) is written with intelligence and elegance, although it requires more background in statistics than this text offers. Less demanding but still interesting is Theodore M. Porter's *Rise of Statistical Thinking: 1820–1900*. F. N. David's *Games, Gods and Gambling* (New York: Oxford, 1962) is a readable, interesting history of probability.

On the importance of playing and thinking, I recommend anything by Douglas Hofstadter both for content and for example. His *Gödel, Escher, Bach* is a tour de force, but you should begin with his essays in *Metamagical Themas* (New York: Basic Books, 1985). Hofstadter plays with ideas with zest and exuberance. To improve your numeracy by learning to handle large numbers, read Douglas Hofstadter's "On Number Numbness" in his *Metamagical Themas*. You will have fun learning to make better sense of federal budgets, McDonald's "Billions and Billions Served," and the rest of the gigantic numbers we live amidst. On numeracy in general, read and learn from John Paulos' *Innumeracy* (New York: Penguin, 1988) and *Beyond Numeracy* (New York: Penguin, 1991).

Most books cited above are not only useful but also quite interesting and readable. But here is my nominee for the dullest statistics book ever published: The Rand Corporation's *Million Random Digits with 100,000 Normal Deviates.* Useful, sure, but very boring.

Index